Social Inequality, Analytical Egalitarianism, and the March Towards Eugenic Explanations in the Social Sciences

Edited by

Laurence S. Moss

WILEY-BLACKWELL

Social Inequality, Analytical Egalitarianism, and the March Towards Eugenic Explanations in the Social Sciences

Edited by

Laurence S. Moss

WILEY-BLACKWELL

This edition first published 2008

Blackwell Publishing was acquired by John Wiley & Sons in February 2007. Blackwell's publishing program has been merged with Wiley's global Scientific, Technical, and Medical business to form Wiley-Blackwell.

Registered Office
John Wiley & Sons Ltd, The Atrium, Southern Gate, Chichester, West Sussex, PO19 8SQ, United Kingdom

Editorial Offices
350 Main Street, Malden, MA 02148-5020, USA
9600 Garsington Road, Oxford, OX4 2DQ, UK
The Atrium, Southern Gate, Chichester, West Sussex, PO19 8SQ, UK

For details of our global editorial offices, for customer services, and for information about how to apply for permission to reuse the copyright material in this book please see our website at www.wiley.com/wiley-blackwell.

Library of Congress Cataloging-in-Publication Data

Social inequality, analytical egalitarianism and the march towards eugenic explanations in the social sciences / edited by Laurence Moss.
p. cm.—(Studies in economic reform and social justice)
ISBN 978-1-4051-9124-1 (softcover)—ISBN 978-1-4051-9125-8 (cased) 1. Peart, Sandra. "Vanity of the philosopher". 2. Economics—History—19th century. 3. Economics—History—20th century. 4. Classical school of economics. 5. Neoclassical school of economics. 6. Equality—Economic aspects. 7. Eugenics—Economic aspects. I. Moss, Laurence S., 1944-

HB85.P43S63 2008
330.15′3—dc22

2008024790

A catalogue record for this book is available from the Library of Congress.

Set in 10 on 13pt Garamond Light by SNP Best-set Typesetter Ltd.
Printed in Singapore by KHL Printing Co Pte Ltd.

01—2008

Contents

Editor's Introduction

This collection is a tribute to the ingenuity and labors of Sandra J. Peart and David M. Levy. In a series of books and articles, this research team (hereafter "PL") has offered the profession something in the nature of a central organizing idea about 19th century economic thought (Levy 2001; Peart and Levy 2005a, 2005b). According to the PL thesis, for most of that century, a clearly identified group of economists adhered to analytical egalitarianism (AE). AE holds that all individuals are equally capable of making economic decisions, despite the fact that the contexts and environments in which these decisions are made vary depending on luck and fortune.

The concept of "equality" that is most vividly portrayed is best brought out by a quotation from the founder of the classical school of economics, Adam Smith. According to Smith, the intellectual differences between a lowly street porter and a philosopher are more the result of custom, chance, and accident than they are the result of any inborn special quality that some might claim to be the result of biological inheritance. The opposite of AE is analytical hierarchy (AH). According to AH, our capacities to experience "utility" vary from individual to individual. Some individuals should (therefore) be preferred or treated better than other individuals because the aggregation of good or intense feelings points us in this direction. As socially minded policy experts, we do not blush about our centrally planned objective of maximizing some measure(s) of overall social welfare. According to PL, AH caught the attention of mainstream economics and gradually took center stage among mainstream economists (Peart and Levy 2005b).

Eventually, economists touched shoulders with the eugenics movements as they unfolded both in Europe and the United States (Moss 2002; Selden 1999). Eugenics is a discipline designed to uncover the causes and conditions under which the local population can be made better or "improved." If certain races of people are desired but are scarce relative to other races, then perhaps a tax rebate is needed for children born to families of the favored race. Singapore has implemented such a tax subsidy to encourage the growth of the Chinese

American Journal of Economics and Sociology, Vol. 67, No. 3 (July, 2008).

cohort of the population. If certain ethnic groups are better noted for their honesty in commerce relative to others, then perhaps a tough immigration policy favoring that ethnic group and tough-minded steps to deport the other, less desired, ethnic group(s) is what is needed. When inspiring public policy, eugenics has a horrific tendency to slip into ideas about euthanasia and ethnic cleansing (Moss 2006). PL underscore the tendency among adherents of AH to slide into the National Socialism of the Hitler period of the 20th century.

These strands of thought are among the most original and stimulating of the last decade or so. Certainly, PL's research has attracted commentary and analysis from this current generation of economic scholars. Most notably, an entire roundtable was devoted to PL's work at the History of Economics Society's Annual Conference held at Grinnell College in June of 2006. These essays are included in this volume, along with a short response by PL. The last essay in this volume, by Terenzio Maccabelli, was not part of this roundtable. The Maccabelli essay serves the much-needed purpose of helping us understand the linkages between the English-language sources and the eugenics literature on the European continent. It is published here for the first time and brings us to that complicated "transition period" at the tail end of the 19th century when (according to PL) AE gave way to AH.

Our first essay, by Charles R. McCann, Jr., reviews the preconditions of a moral order in which hierarchical thinking paves the ground for the racial politics of the 19th century and after. McCann agrees with PL that 19th century economic thought is characterized by a split in thinking about the nature of man. Next, Andrew Farrant provides a more detailed examination of the connecting tissue linking economics and eugenics. Farrant identifies some of what they had in common; for instance, both embraced central planning that aimed to produce measurable demographic results.

Historians of economic thought have done a remarkably good job of carving out of the complex tapestry of 19th century intellectual thought a literature called "economic thought." If discussions about economic phenomena have a tendency toward racial politics, then historians would be compelled to take a second look at which strands of thinking in related areas need to reexamined in order for the story about how economics developed to have greater coherence. PL have not nor cannot ignore the role that religious ideas have played and continue to play in these debates. In his essay, J. Daniel Hammond offers us such a gloss.

The key to unraveling the interaction(s) with religious ideas is to be found in the discussion of the role sympathy plays in human interaction (see Peart and Levy 2005a). It turns out that Catholic doctrine has also made a positive contribution, especially in its promulgation of the doctrine of the brotherhood of the entire human race. The idea that we are close relatives, made in the image of God, damages the foundations of hierarchical thinking in economics. The expert policy maker is not superior to those who are subject to his or her policy decisions and pronouncements.

Hammond criticizes PL when they claim a sharp contrast between evangelical thinking and Papal dicta. Hammond points to the many variations of Protestantism that competed for the loyalty of the English in the centuries following the break with Rome in the 16th century. They may have all been "English" in spirit, but in the application of religious thinking they have had vastly different impacts on how they received biological thinking in policy-making circles. There is no evidence to support the claim that Protestant variants insulated the moral order from racial politics any better than did Catholic thought. Hammond questions the claim that an important alliance existed between classical economics and the mid-19th evangelicals (see also Persky's essay).

Kevin D. Hoover and Joseph Persky criticize PL from different although complementary directions. Hoover documents the several key places where PL confuse the important distinction between analytic egalitarianism and substantive egalitarianism. This "equivocation between analytical and substantive egalitarianism does much of the work" in generating the sharp conclusions PL reach (see, for example, Peart and Levy 2005a). According to Hoover, the logical connections that PL offer are not convincing and must be strengthened.

Persky questions the credibility of PL's claim that classical economics "lost its soul" when it dropped the absolute equality ideas of AE and accepted AH in its policy-making strictures. From a strictly logical or analytic point of view, AH leads to a policy-making stand favoring the redistribution of wealth and income to raise social welfare. The march to eugenic thinking may just have been a bizarre sideshow event.

Next, PL offer a short rebuttal to their several critics. While graciously acknowledging the value and quality of the arguments presented, PL hold to their claim that "eugenics is the oldest form of central planning" and that in the 19th century, AE failed to protect economics from the ravages of racial politics. Here, they place their emphasis on the importance of the idea that the one who models society is at the same time a member of that same society that he or she models. Such an insight holds the key to a variant of equality that may protect the wealthy from the anger and envy of the poor expressed by way of the politics of redistribution. It also holds the key to a variant of equality that flies in the face of the racial politics and ethnic cleansings of the 20th and now 21st centuries.

In his contribution, Farrant reminds us of the connections that exist not only between John Stuart Mill and AE but also those that were established by Mill's father, James Mill. While British associationist psychology may be consistent with AE methods for explaining social phenomena, this approach cannot be held solely responsible for the rise of AH and the eugenics literature at the end of the 19th century. The term "eugenics" as well as the claim that population policy can be rooted in science can both be attributed to the English polymath Francis Galton. Galton noted how selective breeding could improve the moral character of the (average) British citizen. This idea is what sparked ideas in the rest of Europe, especially in France and Germany.

In Terenzio Maccabelli's essay, Georges Vacher de Lapouge and Otto Ammon founded a school of thought that they termed "social anthropology" or "anthropo-sociology." In its starkest terms it held that social inequality is (mostly) caused by racial mechanisms. Translated portions of their writings reached the pages of the *Quarterly Journal of Economics* and the *Journal of Political Economy* through the actions of a young University of Chicago economist, Carlos C. Closson. Closson translated portions of Lapouge's and Ammon's writings into English. After 1900, Closson's name "suddenly vanished from any social and economic debates." He seems to have lost interest in economics and sociology. Maccabelli does his best to gather what little is known about Closson's life and his efforts to promote eugenics in scholarly circles.

Maccabelli also points to the influence of Lapouge and Ammon on Vilfredo Pareto. Pareto's concept of social hierarchies was indebted to eugenic ideas about social heterogeneity. Ultimately, Pareto did not support eugenic explanations of the shape of income distributions. Ammon and Lapouge advocated state intervention to eliminate the "unfit," and thus paved the road to National Socialism in Europe. In the United States, Thorstein Veblen seems to have adopted some of the racial ideas of Lapouge and Ammon. (Veblen was Closson's colleague at the University of Chicago). The story is complex, and we are grateful to PL for opening this area of research to students of social science. This volume of essays extends our discussion of PL's pioneering work linking the history of economics to the central state planning programs of the last century.

References

Levy, David. (2001). *How the Dismal Science Got Its Name.* Ann Arbor: University of Michigan.

Moss, Laurence S. (2002). "Film Note of Peter Cohen's Homo Sapiens 1900." *American Journal of Economics and Sociology* 61 (October): 863–867.

——. (2006). "Ethnic Conflict and the Economics of Social Cooperation: Reflections on a Difficult Problem." In *Money and Markets: Essays in Honor of Leland B. Yeager.* Ed. R. Koppl. London: Routledge.

Peart, Sandra J., and David M. Levy. (2005a). *The 'Vanity of the Philosopher': From Equality to Hierarchy in Postclassical Economics.* Ann Arbor: University of Michigan Press.

——. (2005b). "From Cardinal to Ordinal Utility Theory: Darwin and Differential Capacity for Happiness." *American Journal of Economics and Sociology* 64 (July): 851–879.

Selden, Steven. (1999). *Inheriting Shame: The Story of Eugenics and Racism in America.* New York: Teachers College Press, Columbia University.

II. On the Peart and Levy Thesis

Observations on *The "Vanity of the Philosopher"*

By Charles R. McCann, Jr.*

Abstract. The purpose of this essay is to identify certain points at which the presentation in *Vanity* appears incomplete or in error, especially as regards the interpretations of Mill and Spencer. It is shown that, while the authors have stated their case well, certain mischaracterizations, misstatements, and omissions seem to allow an alternative overall interpretation. However, another look suggests that, despite these discrepancies, the given interpretation is quite valid.

I

Introduction

Sandra Peart and David Levy's *The "Vanity of the Philosopher"* is a book long overdue. It is telling that the story presented will be considered by some to be controversial, and will even be challenged as revisionist or as a piece of propaganda. Such is the nature of the climate within which this debate is to unfold. For too long the Left (of both the academic and coffeehouse varieties) have condemned the market economy and the economists whom they view as its unholy defenders as complicit in an evil endeavor that has as its goal the subjugation of what they contend are the lower classes and the lesser races. Yet when one dares point out the actual details of the development of classical economic thought, as formulated from the late 18th to the mid-19th century, a story that is completely at odds with the accepted fable, accusations are leveled of lack of scholarship, ignorance of the material and the period, inattention to standard

*Charles R. McCann, Jr. is at the Department of Economics, University of Pittsburgh. He is the author of *Individualism and the Social Order: The Social Element in Liberal Thought* (2004).

American Journal of Economics and Sociology, Vol. 67, No. 3 (July, 2008).

references, political bias, cultural insensitivity, and so on. One even hears such absurdities as the following: While the literary types who poisoned the debate with racial epithets as they sought to maintain the status quo of inequality and separateness may have advanced hierarchical views, they actually in the end opposed slavery and so their outbursts may be forgiven.

The authors aim at a more accurate portrayal of events, one not infected with the ideological biases evident in much that passes for scholarship on this topic. The result is a bold, take-no-prisoners defense of the market order and its principal philosophers, Adam Smith and John Stuart Mill. Both argued for equality and human homogeneity as preconditions of a market-oriented social structure, confronting directly the prevailing attitudes of some of the leading intellectuals of the period, and in so doing changing the nature of the political and social debates.[1] The contrast with the later, post-classical economists is indeed striking, and has been distressingly little acknowledged. In fact, one can delve into the literature on such luminaries of political economy as William Stanley Jevons and Francis Ysidro Edgeworth and be hard-pressed to discover any reference to their racial politics. Yet here one is led to see that political and social views are often intertwined with philosophical stances that themselves undergird economic method.

The purpose of this brief paper is not to restate the arguments presented in this compelling tale, as the evidence accumulated is quite impressive, but rather to expand upon specific areas in which the coverage seems somewhat superficial and to identify points at which the presentation appears incomplete or (dare I say) in error. That such comments are brief is a testament to the scholarship of the authors and the thoroughness of their effort.

II

Hume, Sympathy, and Race

THE MAIN ARGUMENT of *Vanity* begins with a consideration of the social philosophy of Adam Smith, briefly contrasted (too briefly, actually) with the views of David Hume. Perhaps this is as it should be, as Smith presents a far more cogent analysis of sympathy, a central theme of the book, than does Hume. In addition, as a representative of the Scottish Enlightenment, Hume presents a particularly distressing example of the influences that may have impacted Smith's philosophical outlook. Fortunately, for the purposes of the present work, his services are useful as a foil of sorts to Smith's own more fully developed position.[2]

The contrast is indeed striking. We are told that "Hume's sympathetic principle moves from *equality in fact* to sympathy" (Peart and Levy 2005: 131, emphasis in original). The *principle* may do so, but does Hume actually himself believe in equality? Peart and Levy correctly point out that Hume's sympathy "is an empathy we feel for those like us," and so Hume is able to ignore those of obvious inferior status or natural ability. This idea of "likeness" is important to keep in mind, for it informs Hume's position with respect to race. Unfortunately, this aspect of Hume's philosophy is not addressed in *Vanity*, even though some mention would have been illustrative of the distinctions the authors wished to make between Hume and Smith. Consider the following, which Peart and Levy do not quote, but that most clearly argues the point:

> I am apt to suspect the negroes, and in general all the other species of men (for there are four or five different kinds) to be naturally inferior to the whites. There never was a civilized nation of any other complexion than white, nor even any individual eminent either in action or speculation. No ingenious manufactures amongst them, no arts, no sciences. On the other hand, the most rude and barbarous of the whites, such as the ancient GERMANS, the present TARTARS, have still something eminent about them, in their valour, form of government, or some other particular. Such a uniform and constant difference could not happen, in so many countries and ages, if nature had not made an original distinction betwixt these breeds of men. Not to mention our colonies, there are NEGROE slaves dispersed all over EUROPE, of which none ever discovered any symptoms of ingenuity; tho' low people, without education, will start up amongst us, and distinguish themselves in every profession. In JAMAICA indeed they talk of one negroe as a man of parts and learning; but 'tis likely he is admired for very slender accomplishments, like a parrot, who speaks a few words plainly. (Hume 1875, Essay XXI: 252n)

From this passage, at least (and it should be mentioned that this is an isolated passage in a footnote to a much later edition of Hume's *Essays Moral, Political, and Literary*), it appears clear that Hume has not only abandoned any pretext of equality, but may actually have held (although the textual evidence is scant indeed) to a belief in polygenesis—how else to reconcile such a presumption that nature made "an original distinction betwixt these breeds of men." Winthrop Jordan, in his magisterial work *White Over Black*, actually cites Hume as an important philosophical influence on 18th- and 19th-century proponents of slavery and racial distinction, especially in the United States; his position even at the time was acknowledged as a "well-known diatribe against the Negro," which was of such authority as to be seen as demanding refutation from anti-slavery advocates (Jordan 1968: 446).

III

Smith and Egalitarianism

THE IMPORTANCE OF Adam Smith in this intellectual tale cannot be overstated, and the authors are to be commended for their insights into this still too-little appreciated aspect of Smith's work. Smith serves to establish the patristic legacy of classical economics—the publication in 1776 of his *Wealth of Nations* is the signal event that defines the beginning of the science.

It is Smith's philosophical position as set out in his 1759 *Theory of Moral Sentiments* that motivates Peart and Levy's story. It is from Smith that classical economics comes to be associated with "an analytical egalitarianism that presumes humans are the same in their capacity for language and trade"; in other words, Smith's moral stance, and therethrough his economic philosophy, presupposes homogeneity (Peart and Levy 2005: 3). This is important because of the implication that any perceived differences among groups (in Smith's view, classes, but one may infer other, including racial, distinctions) can be explained by other than heritability, or a natural distinctness. Such differences as may be observed, for example, "between a philosopher and a common street porter" are more the result of "habit, custom, and education" than of heredity, and so human perfectibility may be achieved through education and social engineering (Smith 1789: I, ii, 4 [15]). (Smith thus appears to adopt a "blank-slate" attitude toward human nature, with the environment being of greater significance than innate ability, and even hints at the social (but obviously not biological) reforms addressed in 19th- and 20th-century British Socialism and American Progressivism. The sociologist Charles Horton Cooley actually studied Smith, and the influence is evident in his writings on the importance of environment in social organization.[3])

In contrast to Hume, Smith *begins* with sympathy and draws from it considerations of equality (Peart and Levy 2005: 131). Sympathy produces feelings of reciprocity and conscience, which manifest in justice and therethrough to generosity (2005: 135). The conclusion is then straightforward: "For Smith, man becomes a moral agent by earning the approbation that comes from recognizing we are all equally deserving of sympathy" (2005: 137). It is through such a recognition that classical economics developed its understanding of human homogeneity, and so structured its postulates upon a moral order emphasizing human equality. Once such recognition is denied, so our authors contend, the way is clear for hierarchical thinking and the racial politics engendered. (Whether homogeneity is a necessary or even sufficient condition for equality is another question, one not addressed here.)

IV

Mill and Paternalism[4]

FROM SMITH we are led to perhaps the preeminent economic and social theorist of the 19th century, John Stuart Mill. Mill, as noted above, is one of the central characters in this drama, continuing after a fashion along Smithian lines while clarifying and even reorienting important conceptual matters. Throughout *Vanity* he is poised as a counter to some rather odious characters such as Thomas Carlyle, John Ruskin, Charles Darwin, and even Francis Edgeworth, as the bulwark against a creeping paternalism, holding fast against the encroachment into philosophy, politics, and economics of the naturalistic fallacy.

Yet while Mill is clearly worthy of such standing, one must accept that there are difficulties with this hagiographic portrayal that the authors are reluctant to consider. One of the nagging questions regarding Mill's essay *On Liberty* is the extent to which he appears to hold to paternalistic notions. Mill's concern in this essay is "to assert one very simple principle, as entitled to govern absolutely the dealings of society with the individual in the way of compulsion and control." This principle, the "liberty principle," maintains

> that the sole end for which mankind are warranted, individually or collectively, in interfering with the liberty of action of any of their number, is self-protection. That the only purpose for which power can be rightfully exercised over any member of a civilised community, against his will, is to prevent harm to others. His own good, either physical or moral, is not a sufficient warrant. (Mill 1859: 13)

Such appears to be a clear and cogent affirmation of the libertarian philosophy. Yet when elaborating on his liberty principle, Mill feels compelled to make certain qualifications. His doctrine "is meant to apply only to human beings in the maturity of their faculties," and so he excludes immediately from consideration children and those below the age of majority. He also removes from consideration the mentally infirm and those under the care of another (Mill 1859: 13). He then goes further, and eliminates from coverage all incapable of "spontaneous progress," a class that includes "those backward states of society in which the race itself may be considered as in its nonage" (1859: 13). His reasoning is interesting:

> Despotism is a legitimate mode of government in dealing with barbarians, provided the end be their improvement, and the means justified by actually effecting that end. Liberty, as a principle, has no application to any state of things anterior to the time when mankind have become capable of being improved by free and equal discussion. Until then, there is nothing for them but implicit obedience to an Akbar or Charlemagne, if they are so fortunate as to find one. (Mill 1859: 13–14)

Thus does Mill appear to accept at least *some* heterogeneity (although such a conclusion is contingent on the meanings of "backward states," "nonage," and "barbarians") as he suggests (or, more accurately, one may, with some license, *infer*) that those uncivilized peoples, being either childlike or deficient, and (perhaps) even innately (racially?) inferior, are ill-suited to the same liberties to which the civilized peoples are entitled; Mill seeks a compromise between environment and heredity. (It should be understood that there is no evidence whatever of racism or racial bias in any of Mill's writings.) In any event, Mill's assertion of the need for some sort of control over those ill-equipped to see to their own welfare, while perhaps an isolated thought regarding the appropriate scope of limitations on individual freedom, is nonetheless instructive in understanding his attitude toward social control. This attitude is potentially more far-reaching than Peart and Levy are willing to admit. One may reasonably conclude that even Mill would not have disagreed completely with the notion that, even among the "civilized," one must take care "to look after systematically poor optimizers whose preferences could not be trusted" (Peart and Levy 2005: 12).

This is not the only place in which Mill deems the application of the liberty principle inoperable. The liberty principle applies in respect of the conduct of *rational* people who have reached the age of majority, conduct that has no external manifestations. (Presumably, then, "barbarians" are nonrational as well as childlike.) There are, however, two glaring exceptions to this rule, areas of great social interest in which "misplaced notions of liberty prevent moral obligations on the part of parents from being recognised, and legal obligations from being imposed" (Mill 1859: 108). These exceptions are education and marriage. It is here in the exceptions that we see the working of the rule, as the externalities to be guarded against are *social*, appertaining to the welfare of the whole.

In the case of compulsory education, Mill's reasoning is illuminating as to the extent of his rule. Each family has a moral duty to provide for the education of the children to ensure that each will "perform his part well in life towards others and towards himself"—to neglect this duty is a "moral crime" (Mill 1859: 105). The state, therefore, is justified in the interests of the whole to see that such an obligation is maintained. While he is opposed to the state *providing* education, Mill is adamant in his desire that the state *require* such provision (1859:105–106).

In the case of marriage and childrearing, Mill's paternalism is much more in evidence. Mill considers it a crime to produce a child unless the child "will have at least the ordinary chances of a desirable existence." In those circumstances of overpopulation, instances in which an increase would result in a greater competition for employment and hence result in a diminution of the share of the product available to the existing labor pool, Mill goes so far as to insist that the production of offspring "beyond a very small number . . . is a serious offence against all who live by the remuneration of their labour" (Mill 1859: 108). His preference is to allow such laws as "forbid marriage unless the parties can show that they have the means of supporting a family," as such laws "do not exceed the legitimate powers of the State," but rather are interferences designed "to prohibit a mischievous act—an act injurious to others, which ought to be a subject of reprobation, and social stigma, even when it is not deemed expedient to superadd legal punishment" (1859: 108).

While in both instances—education of children and marriage and childrearing—Mill's argument for interference may be interpreted as conducive to the social good, suggesting an aggregated Utilitarianism, his liberty principle as he himself defines it is directed against *all* state interferences with self-regarding actions, and so the notion of furthering the social good would appear antithetical to a commitment to individual liberty. In most instances it is sufficient that the individual be apprised of the consequences of a considered act, such that he or she is able to make an informed decision. This is what has been termed "soft-paternalism" (Riley 1998: 198). The justification for allowing state interference in such areas as Mill identifies is that, in these specific instances—education and childrearing—the consequences are inherently other-directed. This fact alone is sufficient for Mill to defer to the will of the state and its "experts" in setting limits to personal conduct.

(As Peart and Levy do not comment on Mill's associationist theory of learning, a challenge to the intuitionist philosophy of the Continental rationalists, we shall do likewise, except to note that this represents a continuation of the age-old nature/nurture debate, with Mill, consistent with Smith, taking the nurture side. Steven Pinker (2002) identifies the fallacies in Mill's position, a version of "blank-slate" psychology.)

V

Darwin, Spencer, and Heterogeneity[5]

Charles Darwin's place in this story should not be minimized, and Peart and Levy do well in according him pride of place in the shift within Victorian social philosophy from homogeneity to heterogeneity. While Darwin articulates a view of the moral sense quite similar to that of Smith, he employs it in a *social*, not an *individual*, sense, and so is far more the communitarian than the classical liberal. Darwin's man is neither *homo economicus* nor *homo ethicus*, but rather *homo socialis*, and so he requires a definition of a standard of social welfare. This he provides in *The Descent of Man*. For Darwin, the general good "may be defined as the rearing of the greatest number of individuals in full vigour and health, with all their faculties perfect, under the conditions to which they are subject" (Darwin 1879: 145). Thus to advance the perfection of the race is to further the general good. That this may be possible through selective breeding, with marriages arranged to the end of the betterment of the social order, Darwin readily accepts (1879: 688).

It is in respect of such statements as these that Darwin was viewed as having given his approval to measures directed at racial betterment, and so *The Descent of Man* is an important volume in the library of the eugenics movement.

The choice of Herbert Spencer as a foil to Darwin is quite interesting, as his philosophical perspective is in many ways reflective of that of the classical economist. The authors note that "Spencer's goal was egalitarian, while that of Darwin entailed biological perfection or hierarchy" (Peart and Levy 2005: 210), and so Spencer is in this sense on the side of the angels. Of course there is much in Spencer to support this conclusion. His interpretation of the moral sense is predominantly Smithian, moral action being (in the more highly developed state) not obligatory or coerced, but internalized and so more in the nature of unconscious desire; the moral sense is that "special agent by which the distinction between right and wrong exercise of faculties is recognised and responded to" (Spencer 1851: 90). Furthermore, one finds a striking similarity to certain particulars in Mill's social and political philosophy. In *Social Statics* (1851), Spencer actually states a liberty principle—his "First Principle"—virtually identical to the one advanced by Mill some eight years later: "Every man has freedom to do all that he wills, provided he infringes not the equal freedom of any other man" (Spencer 1851: 103).

While Spencer indeed requires sympathy and egalitarianism as the bases of his utilitarian philosophy, and so has affinities to Smith and Mill, in his theory of evolution one sees distinctions emerge. First,

Spencer's evolutionary process is teleological: Social progress is a continuous movement to a more complex, or more perfect, form. In addition, and more to the point, while an ardent believer in monogenism, his belief in racial distinctions compels him to advance a theory to explain such differences. In *First Principles* he explicitly states that the path of development of a species is from homogeneity to heterogeneity, with racial differentiation being one end result:

> Whether an advance from the homogeneous to the heterogeneous is or is not displayed in the biological history of the globe, it is clearly enough displayed in the progress of the latest and most heterogeneous creature—Man. It is alike true that, during the period in which the Earth has been peopled, the human organism has grown more heterogeneous among the civilized divisions of the species; and that the species, as a whole, has been made more heterogeneous by the multiplication of races and the differentiation of these races from each other. (Spencer 1894: Ch.15, §121)

Such racial heterogeneity would be the obvious result even if one believed in polygenism:

> Even were we to admit the hypothesis that Mankind originated from several separate stocks, it would still remain true that as, from each of these stocks, there have sprung many now widely different tribes, which are proved by philological evidence to have had a common origin, the race as a whole is far less homogeneous that it once was. (Spencer 1894: Ch.15, §121)

Again, in *The Principles of Biology*, Spencer's position on evolutionary development is made quite clear: Heterogeneity is a natural evolutionary process, the biological equivalent of entropy:

> Unless we deny the persistence of force, we must admit that the gravitation of an organism's structure from an indefinitely homogeneous to a definitely heterogeneous state, must be cumulative in successive generations, if the forces causing it continue to act. And for the like reasons, the increasing assemblage of individuals arising from a common stock, is also liable to lose its original uniformity; and, in successive generations, to grow more pronounced in its multiformity. (Spencer 1895: Ch.10, §158)

Finally, as if any additional confirmation were needed, we should mention the following from *The Principles of Sociology*, which serves as witness to Spencer's notion of racial hierarchy:

> Large areas such as parts of Africa present, which prove fatal to the higher races of men, and the steaming atmospheres of which cause enervation,

may continue to be inhabited by lower races of men, subject to political arrangements adapted to them. (Spencer 1912: Vol. II, §577)

One may conclude, then, that Spencer's social philosophy is, despite superficial similarities, not of the Smithian variety: Increasing complexity and heterogeneity are the hallmarks of Spencerian evolutionary development; homogeneity at best serves as a point of departure (giving Spencer's theory of evolution, biological as well as social, a "big bang" quality). Yet one may as well conclude that, while Spencer's work is much less prone to abuse than is Darwin's, as Spencer does not allow for the eugenic conclusions that may be seen to be an obvious consequence of Darwinian social biology (such as can be derived from *The Descent of Man*), one should be mindful of the road to which Spencerian conclusions may lead.

VI

Remarks

WHILE SOME CONSIDERATION should be allowed for the above-listed references, they appear to be qualifications to the overall argument of Peart and Levy, emendations that, in a sense, provide the exceptions that serve to prove the rule. The overall thesis and conclusions as to the disconnect between the classicals and the neoclassicals respecting the nature of man remain valid.

All in all, *Vanity* is a precautionary tale, one that deserves a wide audience. Further, it is my sincere hope that this effort will generate interest in the topics addressed and that more critical analyses may be forthcoming.

Notes

1. Nonetheless, despite overwhelming evidence to the contrary, as presented both here and in Levy's previous book on the subject (Levy 2001), it has been claimed, by, for example, Susan Zlotnick, that Smith and Mill were racists, since their writings advanced the notion of the superiority of Western civilization. If for no other reason than to counter such baseless contentions, Peart and Levy is essential reading.

2. The psychologist Marc Hauser notes that Hume's works "are now mandatory reading at most universities," and that his "ideas have enjoyed a rebirth in the wake of new developments in the mind sciences" (Hauser 2006: 23).

3. On the influence of Smith on Cooley, see Jacobs (2006). On Cooley's social philosophy, see McCann (2006).

4. A more extensive presentation of Mill's social and political philosophy may be found in McCann (2004).

5. Spencer's socioethical position is detailed in McCann (2004).

References

Darwin, Charles. ([1879] 2004). *The Descent of Man, and Selection in Relation to Sex*. London: Penguin. (First edition: 1871.)

Hauser, Marc D. (2006). *Moral Minds: How Nature Designed Our Universal Sense of Right and Wrong*. New York: Ecco/HarperCollins.

Hume, David. (1875). *Essays Moral, Political, and Literary*. Eds. T. H. Green and T. H. Grose. London: Longmans, Green and Co.

Jacobs, Glenn. (2006). *Charles Horton Cooley: Imagining Social Reality*. Amherst: University of Massachusetts Press.

Jordan, Winthrop D. (1968). *White Over Black: American Attitudes Toward the Negro, 1550–1812*. Chapel Hill: University of North Carolina Press.

Levy, David M. (2001). *How the Dismal Science Got Its Name: Classical Economics and the Ur-Text of Racial Politics*. Ann Arbor: University of Michigan Press.

McCann, C. R., Jr. (2004). *Individualism and the Social Order: The Social Element in Liberal Thought*. London: Routledge.

——. (2006). "Degeneracy, Eugenics, and Directed Order: Charles Horton Cooley and American Progressivism." Paper presented at the tenth annual meeting of the European Society for the History of Economic Thought, University of Porto, Portugal, 28–30 April.

Mill, John Stuart. ([1859] 1989). "On Liberty." In *On Liberty, and Other Writings*. Ed. Stefan Collini. Cambridge: Cambridge University Press.

Peart, Sandra, and David M. Levy. (2005). *The "Vanity of the Philosopher": From Equality to Hierarchy in Post-Classical Economics*. Ann Arbor: University of Michigan Press.

Pinker, Steven. (2002). *The Blank Slate: The Modern Denial of Human Nature*. New York: Viking.

Riley, Jonathan. (1998) *Routledge Philosophy Guidebook to Mill, on Liberty*. London: Routledge.

Smith, Adam. (1789 [1937]). *An Inquiry into the Nature and Causes of the Wealth of Nations*, 5th ed. Ed. Edwin Cannan. New York: Modern Library. (First edition: 1776.)

Spencer, Herbert. (1851). *Social Statics: or, The Conditions Essential to Human Happiness Specified, and the First of Them Developed*. London: John Chapman.

——. (1894). *First Principles*. New York: D. Appleton and Co.

——. (1895). *The Principles of Biology.* New York: D. Appleton and Co.
——. (1912). *The Principles of Sociology.* New York: D. Appleton and Co.
Zlotnick, Susan. (2004). "Contextualizing David Levy's *How the Dismal Science Got Its Name*; or, Revisiting the Victorian Context of David Levy's History of Race and Economics." In *Race, Liberalism, and Economics.* Eds. David Colander, Robert E. Prasch, and Falgugi A. Sheth. Ann Arbor: University of Michigan Press.

The "Vanity of the Philosopher"

Analytical Egalitarianism, Associationist Psychology, and Eugenic Remaking?

By ANDREW FARRANT*

ABSTRACT. These comments explore the relationship between analytical egalitarianism, race-blind theorizing, and associationist psychology. Associationist psychology, though making an implicit appearance in *Vanity*, was central to the egalitarian analysis provided by James Mill and John Stuart Mill. Indeed, associationist considerations lay at the heart of Mill's race-blind analysis of cottier tenure, and his exchange with Thomas Carlyle over the "Negro Question." These comments also note some intriguing comparisons between the debate over eugenics and the debate over socialist calculation.

1

Introduction

THE "VANITY OF THE PHILOSOPHER" (Peart and Levy 2005) provides a welcome addition to the burgeoning literature on the role that racial themes and eugenic doctrine played in the development of late 19th- and early 20th-century social science (see, e.g., Colander, Prasch, and Sheth 2004; Cot 2005; Leonard 2005a, 2005b; Levy 2001). Peart and Levy's mastery of the texts is apparent throughout *Vanity*, and they ably muster a truly impressive array of evidence to bolster their thesis that the rise of "postclassical economics" entailed the "loss of sympathy in economic analysis, and the endorsement of eugenical remaking" by a bevy of late 19th- and early 20th-century social scientists (Peart and Levy 2005: xi).

*Andrew Farrant is at the Department of Economics, Dickinson College, Carlisle, PA, 17013; e-mail: farranta@dickinson.edu. He has research interests in political economy and social philosophy. Thanks are due to Larry Moss for his encouragement in putting the original panel on *Vanity* together and assistance and advice.

American Journal of Economics and Sociology, Vol. 67, No. 3 (July, 2008).

Rather than providing an in-depth review of *Vanity*, my comments here primarily focus on a puzzling omission from *Vanity*'s narrative: associationist psychology. This is a puzzling omission because the tenets of associationist psychology play a vitally important role in both John Stuart Mill's race-blind analysis of cottier tenure in Ireland and Mill's exchange with Thomas Carlyle over the "Negro Question" (Mill 1850).[1] I conclude by noting a rather intriguing complementarity between the 1930s debate over eugenic remaking and the 1930s debate over socialist economic calculation. Additionally, I provide some suggestions for future research along the lines of *Vanity*. For example, why did F. A. Hayek never attack eugenic planning? First, however, we turn to associationist psychology and its intimate connection to Mill's race-blind political economy.

II

Egalitarianism?

PEART AND LEVY intriguingly note that many colleagues (including James M. Buchanan) have asked: "What if Plato were right about inherent differences among peoples?" (Peart and Levy 2005: 125). Plato, of course, exemplifies the hieararchicalist view that supposedly innate differences explain observed heterogeneity. Indeed, Plato's account of the division of labor presupposes innate differences among people.[2] Adam Smith's account of the division of labor, by contrast, contends that people are basically alike at birth, and that it is the division of labor—learning by doing—that induces any later observed heterogeneity in agents' productive capacities. My guess is that Peart and Levy's egalitarianism would not be swayed one iota by any affirmative answer to the Plato question. Indeed, Peart and Levy make abundantly clear that while they do think people "*are, in fact,* different" (2005: xv) on various physical margins, they readily subscribe to Lionel Robbins's view that all systems of "political calculations which do not treat . . . [people] *as if* they were equal are morally revolting" (Lionel Robbins, qtd. in Peart and Levy 2005: xv).[3] Intriguingly, Peart and Levy state that they wholeheartedly subscribe to the analytical egalitarianism of classical economics that, "[s]tarting with Adam Smith . . . presumes humans are the same in their capacity for language and trade; observed differences are then explained by incentives, luck, and history" (2005: 3). Peart and Levy's reference to Robbins (2005: xv) is particularly intriguing because Robbins places much emphasis on the difference between Plato's and Smith's rival accounts of the division of labor in his justly famous LSE lectures (Robbins 1998). Intriguingly, although Robbins argued that "nearly everybody who talks about differences of nature, including professional natural scientists, talks nonsense, and nonsense which can very easily be converted into immoral acts," he did consider it "important to realise that *Socrates had hit on something* . . . There are differences of nature, although we don't at this moment know awfully much about them" (Robbins 1998: 14, emphasis added). Similarly, Robbins suggests that Adam Smith's account of the division of labor—"we all come into the world with, roughly speaking, equal potentiality, and what happens afterwards is a matter of education and experience"—is not "exactly right" (Robbins 1998: 130). Ultimately, however, Robbins is characteristically circumspect: "Don't think that I want to thrust down your throats the *genetic differences*, about which we know very, very little indeed, and most people who talk about them are frauds" (Robbins 1998: 130, emphasis added).[4] Thus I wonder if Peart and Levy are rather closer to Robbins's apparent advocacy of *as if*—rather than substantive—egalitarianism than they may have thought.

As Robbins notes, Smith considers nurture (habit, custom, and education) to trump any natural differences:

> The *difference* between the most dissimilar characters, between a philosopher and a common street porter, for example, *seems to arise not so much from nature, as from habit, custom, and education.* When they came into the world, and for the first six or eight years of their existence, they were, perhaps, very much alike, and neither their parents nor playfellows could perceive any remarkable difference. About that age, or soon after, they come to be employed in very different occupations. (Smith [1776] 1981: 28–29, emphasis added)

At which juncture, "[t]*he difference of talents comes then to be taken notice of, and widens by degrees, till at last the vanity of the philosopher is willing to acknowledge scarce any resemblance.* But without the disposition to truck, barter, and exchange, every man must have procured to himself every necessary and conveniency of life which he wanted. *All must have had the same duties to perform, and the same work to do, and there could have been no such difference of employment as could alone give occasion to any great difference of talents"* (Smith [1776] 1981: 29, emphasis added). Accordingly, heterogeneity is the child of circumstance rather than of nature.

Similar views pervade J. S. Mill's account of the division of labor. Though considering "natural differences . . . something," Mill contends that "*habit is much more*" (Mill [1848] 1965: 127, emphasis added). Indeed, Mill argues that "habit . . . much more than nature" explains why women are apparently more adept than men at turning their hands to a wide variety of tasks ([1848] 1965: 127): "The occupations of nine out of every ten men are special, those of nine out of ten women general, embracing a multitude of details, each of which requires very little time. Women are in the *constant practice* of passing quickly from one manual, and still more from one mental operation to another . . . while a man's occupation generally consists in working steadily for a long time at one thing, or one very limited class of things" ([1848] 1965: 128, emphasis added). The situations, however, "are sometimes reversed, *and with them the characters*" ([1848] 1965: 128, emphasis added). Indeed, "women are not found less efficient than men for the uniformity of factory work" ([1848] 1965: 128). Similarly, when contrasting the habitual behavior of the Irish cottier tenant and the French peasant proprietor, Mill remarks that of "all tillers of the soil, the cottier is the one who has the least to gain by any voluntary exertion; the small proprietor has most . . . [This] is but the natural result of their circumstances. *Put each in the situation of the other, and their characters will be reversed*" (Mill 1986: 958). Mill's remarks about habit, natural differences, and the division of labor, however, together with his statements about the behavior of the Irish cottier and peasant proprietor, fit squarely with the tenets of associationist psychology.[5] For the associationist, "[t]he mind is a blank slate in which experiences are conjoined together to form pleasurable or painful associations" (Sheth 2004: 107). Associationist doctrine contended that the "character of the human mind consists in the sequences of its ideas" (James Mill 1992: 151), and these sequences of ideas—mental trains of association (whether good trains or bad)—terminate in the "idea of some future pleasure or pain" (James Mill 1992: 153). Pleasure or pain hence becomes habitually associated with particular types of behavior.[6]

III

Associationist Psychology and the Architecture of Character?

As noted earlier, "associationist psychology" does not appear in *Vanity*'s pages. Peart and Levy, however, implicitly recognize the importance that trains of association play in Mill's analysis of "human development . . . in the context of institutional change . . . Material desires may be helpful for educational purposes in the transition [from slavery] to freedom" (Peart and Levy 2005: 171). As Mill puts it, while "minds are coarse they require coarse stimuli, and let them have them" (Mill [1848] 1965: 754).[7]

Though a rather arcane topic, associationist psychology is bubbling away under the surface in many of the debates that Peart and Levy study. For example, Mill's pointed strictures against Carlyle's supposition of innate Negro inferiority provide a classic example of associationist thinking.[8] Similar considerations pervade Mill's race-blind analysis of cottier tenure in Ireland.[9]

Intriguingly Peart and Levy note F. Y. Edgeworth's attack on the Benthamite view that "all men, if not equal, are at least *equipotential*, in virtue of equal educability" (Edgeworth, qtd. in Peart and Levy 2005: 8). Again, however, associationist psychology rears its head. James Mill's *Essay on Education* is much to the point: Education denotes "every thing, which acts upon the being as it comes from the hand of nature, in such a manner as to modify the mind" (James Mill 1992: 159–160).[10] Exposure to different sequences of ideas (trains of association) induces heterogeneity in character and conduct: "[L]arge numbers of bodies of men are raised to a high degree of mental excellence . . . Other large bodies, or whole nations, have been found in so very low a mental state, as to be little above the brutes. *All this vast distance is undeniably the effect of education*" (1992: 161, emphasis added).[11] Potential equal competence is rather more than an analytical presumption for adherents to the tenets of associationist psychology: J. S. Mill meant what he said in his *Autobiography* about his achievements being all due to the education he received at the hands of his father rather than any innate ability (Robbins 1998: 220).

Peart and Levy explore the connection between sympathy and the development of character in Chapters 2 and 9 of *Vanity*. Once again, however, considerations of associationist psychology are relevant. Both James and John Stuart Mill thought that education (always broadly understood) had the potential to induce greater sympathy.[12] Similarly, repeated exposure to "bad trains of association" could narrow sympathy or, ultimately, crowd out sympathy altogether: "a human being, almost constantly in pain, hardly visited by a single pleasure, and almost shut out from hope, *loses by degrees all sympathy with his fellow creatures*; contracts even a jealousy of their pleasures, and at last a hatred; and would like to see all the rest of mankind as wretched as himself" (James Mill 1992: 171–172, emphasis added).[13]

IV

Eugenics and Socialist Calculation?

I TURN NOW, however, to Chapter 6 of *Vanity* ("Picking Losers for Sterilization"). Peart and Levy situate eugenics (demographic central planning) within the debate over central economic planning more generally understood (the interwar socialist calculation debate). Mises and Hayek are famous for having vehemently opposed economic planning (the planning of prices and material output) during the 1930s and 1940s. Peart and Levy ask the fascinating "natural follow-up question: were the opponents of material forms of central planning also opposed to planning for the quality of human beings?" (2005: 105). Their narrative focuses on Karl Popper's opposition to Plato's advocacy of eugenics. Hayek never wrote on eugenics to my knowledge, though he did play an instrumental role in the publication of Popper's *The Open Society* (Peart and Levy 2005: 105). Ludwig von Mises, however, provided a characteristically scathing opposition to Nazi eugenic planning:

> The Nazi plan was more comprehensive and therefore more pernicious than that of the Marxians. It aimed at abolishing *laisser-faire* not only in the production of material goods, but no less in the production of men. The Fuhrer was not only the general manager of all industries; he was also the general manager of the breeding-farm intent upon rearing superior men and eliminating inferior stock. A grandiose scheme of eugenics was to be put into effect according to "scientific" principles. (Mises 1981: 531)

Moreover, it is

> vain for the champions of eugenics to protest that they did not mean what the Nazis executed. Eugenics aims at placing some men, backed by the police power, in complete control of human reproduction. It suggests that the methods applied to domestic animals be applied to men. This is precisely what the Nazis tried to do. The only objection which a consistent eugenist can raise is that his own plan differs from that of the Nazi scholars and that he wants to rear another type of men than the Nazis. As every supporter of economic planning aims at the execution of his own plan only, so every advocate of eugenic planning aims at the execution of his own plan and wants himself to act as the breeder of human stock. (Mises 1981: 531)

Rather intriguingly, Peart and Levy wonder about the existence of "that oddity . . . the laissez-faire eugenicist" (2005: 81). I conjecture that John R. Baker might adequately fit the bill. Baker, an early member of the Mont Pelerin Society, was a keen admirer of Hayek's *Road to Serfdom* and a co-founder (with Michael Polanyi) of the Society for Freedom in Science (Kenny 2004). More importantly, however, Baker argued that the welfare state was dysgenic (Kenny 2004). Interestingly, Hartwell (1995: 229) reports that Baker apparently resigned from the Mont Pelerin Society on the grounds that as a noneconomist there was relatively little role for him to play in the Society's work.

Peart and Levy rightly note the difference between the tenor of Popper's attack on eugenic doctrine and Hayek's attack on economic planning. Popper impugned the motives of his opponents: "[B]ehind the sovereignty of the philosopher king stands the quest for power. The beautiful portrait of the sovereign is a self-portrait [of Plato]" (Popper, qtd. in Peart and Levy 2005: 106). Hayek, however, took pains to never impugn the motives of his opponents (Hayek [1949] 1967: 184). Indeed, Schumpeter famously chided Hayek for never attributing anything other than intellectual error to the advocates of economic planning (Schumpeter 1946: 269). Peart and Levy's intriguing juxtaposition of Hayek and Popper, together with their "natural follow-up question," leads me to offer the following speculations.

The participants in the great socialist calculation debate of the 1930s and 1940s—Hayek and the market socialists alike—were extremely loath to discuss any questions relating to planner motivation (Levy 1990). Indeed, Hayek remarked that the economist necessarily had to remain silent whenever the question was raised as to "whether men in general could be trusted to have the moral and psychological qualities which were dimly seen to be essential if a socialist system were to work" (Hayek 1935: 2). Hayek, however, was not the only participant in the planning debate to shy away from such issues. Oskar Lange, for example, rather famously stated that any qualms about the incentive-compatibility of economic planning were the subject matter of sociology rather than economics (Lange 1936–1937: 127). Abba Lerner took pretty much the same position on any questions relating to incentives and planning (Lerner 1937: 267). In retrospect, the pains that the leading protagonists in the calculation debate took to evade any discussion of incentives-related questions are truly remarkable. This, however, is where my speculation rears its head. Were such evasions largely reflective of a desire to avoid getting dragged into a controversy about eugenic remaking and the malleability of human nature—the new "public-spirited" socialist man and the suchlike—in general?

Rather intriguingly, Hayek repeatedly took various "men of science" (advocates of central economic planning to a man) to task during the 1930s and 1940s for their apparent ignorance of economics (see, e.g., Hayek [1941] 1997: 213–220). The only one of these fascinating characters to feature in Peart and Levy's narrative, however, is Lancelot Hogben (the first and only Professor of Social Biology at the LSE). Peart and Levy rightly pay attention to Hogben because he recognized the asymmetric incentives that Peart and Levy argue are operative in eugenic science (Peart and Levy 2005: 122–124). Peart and Levy aptly note that "from the point of view of the eugenic researcher, eugenic research is a benefit and not a cost. . . . Answers of one sort, but not of another, provide power and plenty" (2005: 124). Hogben, however, did not recognize that a similar asymmetry may be operative in the context of economic planning (see Levy 1990). Though Hayek referred to Hogben as an "authoritarian socialist" (Hayek [1940] 1997: 139), Hogben provided a spirited and very public opposition to eugenic doctrine throughout the 1930s and 1940s (Peart and Levy 2005: 122–124).[14]

All of this, however, leads me to wonder whether (Hogben aside, of course) the economic planning finding favor with Hayek's "men of science" was merely a subsidiary part of a rather grander scheme for eugenic remaking? Interestingly, the only mention I have found of eugenics in the 1930s and 1940s literature on socialist calculation—Mises's colorful strictures against eugenics aside—is provided by H. D. Dickinson (1939). Dickinson remarks that:

> [the] qualitative aspect of the population problem is . . . knotty. In the present stage of social biology our knowledge of the innate qualities of different human stocks is so scanty and uncertain that there is little scope for any measures of discrimination between potential parents. (Dickinson 1939: 147)

Dickinson notes that some

> defects are known to be hereditary, and persons who are genetic carriers of these defects should be encouraged to refrain from parenthood. If necessary they must be segregated or sterilized. If their unsuitability for parenthood is biological not social . . . they should be allowed to adopt children . . . [B]eyond such measures of negative eugenics there is little that can be done in the present state of our scientific knowledge. Until one or two generations have grown up in good physical and social surroundings, with equal opportunities of good food, shelter, and education for all, it is impossible to distinguish between the effects of environment and those of heredity. (1939: 147)

Intriguingly, Dickinson cites Hogben at this juncture.[15] To my mind, this raises the question as to why Hayek—ever-ready to do battle with any other variety of planning—never directly attacked eugenics. One relatively plausible possibility is that Hayek was wary of getting involved in a controversy that could all too easily cause him some intellectual embarrassment. For one thing, we can readily imagine the "men of science" suggesting that Hayek lacked the expertise to pontificate on eugenic science (the very charge, after all, that he made with regard to their pronouncements on economics).[16] More research is clearly warranted.

Notes

1. Associationism was for Mill what he said Comte's version of history was for the French thinker, " 'the key to [his] other generalizations . . . the

backbone of his philosophy' . . . It was one of those 'ultimate truths' upon which other truths depend" (Carlisle 1991: 20).

2. For instance, "should the farmer provide enough food for all . . . and devote enough time and labour to food production to provide for the needs of all four [?] Or . . . should he disregard the others, and devote a quarter of his time to producing a quarter the amount of food, and the other three quarters one to building himself a house, one to making clothes, and another to making shoes? Should he . . . devote himself to providing for his own needs only?" (Plato 1987: 59). "[A]s you were speaking, *it occurred to me that in the first place, no two of us are born exactly alike. We have different natural aptitudes, which fit us for different jobs*" (1987: 59, emphasis added). "Quantity and quality are therefore more easily produced when a man specializes appropriately on a single job *for which he is naturally fitted*, and neglects all others" (1987: 60, emphasis added).

3. Peart and Levy rather wryly observe that while one co-author is five feet two inches and the other is six feet tall in height, "relative price changes are not likely to reduce this difference" (Peart and Levy 2005: xv).

4. In response to Smith's famous remarks about the philosopher and common street porter, Robbins states: "[W]hen I think of the difference between me and Einstein, I'm not at all sure that he's right, and I don't think that Einstein's genius, or the genius of Newton and many other people, arose simply from circumstances and education, although no doubt it played a part" (Robbins 1998: 134).

5. J. S. Mill remarked that in "psychology, his [father's] fundamental doctrine was the formation of all human character by circumstances, through the universal Principle of Association, and the consequent unlimited possibility of improving the moral and intellectual condition of mankind by education. *Of all his doctrines none was more important than this, or needs more to be insisted on*" (J. S. Mill 1989: 95–96, emphasis added).

6. "In his *Analysis of the Phenomena of the Human Mind*, James Mill tried to describe in detail the way in which, as a result of associative mechanisms, men can gradually develop a love for humanity which will thereafter direct their conduct" (Passmore 2000: 270).

7. "To civilize a savage, he must be inspired with new wants and desires . . . provided that their gratification can be a motive to steady and regular bodily and mental exertion. If the negroes of Jamaica and Demerara, after their emancipation, had contented themselves . . . with the necessaries of life . . . and abandoned all labour beyond the little which in a tropical climate . . . is sufficient to support existence . . . they would have sunk into a condition more barbarous, though less unhappy, than their previous state of slavery" (Mill, qtd. in Peart and Levy 2005: 171). Similarly, Mill remarks: "In every plan which has to do with Irish labourers the obstacle is—too little ambition, too few wants; the danger . . . [they] will prefer rags, a turf cabin,

and idleness, to comfort with labour. When a people of this character is discovered in the South Sea islands or on the African coast, our merchants . . . excite new wants by placing suitable objects before their eyes. . . . The [Irish peasantry] . . . must have objects before their eyes sufficiently attractive to be worth working for. Comforts and conveniences must be shown to them—must be brought within easy reach" (Mill 1986: 1026).

8. Mill charges Carlyle with falling prey to the "vulgar error" of "imputing every difference which he finds among human beings to an original difference of nature. As well it might be said, that of two trees, sprung from the same stock, one cannot be taller than another but from greater vigour in the original seedling. *Is nothing to be attributed to soil, nothing to climate, nothing to difference of exposure—has no storm swept over the one and not the other, no lightning scathed it, no beast browsed on it, no passing stranger stript off its leaves or its bark?*" (Mill 1850, 29; emphasis added).

9. Mill wrote: "Is it not, then, a bitter satire on the mode in which opinions are formed on the most important problems of human nature and life, to find public instructors of the greatest pretension, imputing the backwardness of Irish industry, and the want of energy of the Irish people in improving their condition, to a peculiar indolence and *insouciance* in the Celtic race? Of all vulgar modes of escaping from the consideration of the effect of social and moral influences on the human mind, the most vulgar is that of attributing the diversities of conduct and character to natural differences" (Mill [1848] 1965: 319). The particular public instructor Mill has in mind is Thomas Campbell Foster—the *Times* correspondent on the agricultural situation in Ireland (see Mill 1986: 887).

10. "[I]t is education wholly which constitutes the remarkable difference between the Turk and the Englishman, and even that still more remarkable difference between the most cultivated European and the wildest savage. Whatever is made of any *class* of men, we may then be sure is possible to be made of the whole human race. What a field for exertion! What a prize to be won!" (James Mill 1992: 147)

11. One can all too readily imagine the chagrin felt by Thomas Carlyle (one of the villains in Peart and Levy's narrative, and rightly so) when coming upon Mill's egalitarian heresy (advocacy of genetic equality): "It is maintained by Helvetius and his set, that an infant of genius is quite the same as any other infant, only that certain surprisingly favourable influences accompany him through life, especially through childhood, and expand him, while others lie close-folded and continue dunces. Herein, say they, consists the whole difference between an inspired Prophet and a double-barrelled Game-preserver. . . . 'With which opinion,' cries Teufelsdrockh, 'I should as soon agree with as with this other, that an acorn might, by favourable or unfavourable influences of soil and climate, be nursed into a cabbage, or the cabbage-seed into an oak' " (Carlyle, qtd. in Levy 2001: 88).

12. "The real effective education of a people is given them by the circumstances by which they are surrounded . . . What shapes the character is not what is purposely taught, so much as the unintentional teaching of institutions and social relations" (J. S. Mill 1986: 955). James Mill's desideratum is one where the "larger associations [a greater degree of sympathy], generated by a good Education, control the narrow associations, growing out of a particular position" (1829: 229). "A very general idea, such as that of *Mankind*, is an indistinct idea; and no strong association is formed with it, except by the means of Education. In the common run of men, the narrow sympathies, alone, act with any considerable force. Such men can sympathize with . . . their own Family, or their own class. But to sympathize with mankind at large, or even with the body of the people in their own country, exceeds the bounds of their contracted affections" (1829: 231–232).

13. "How few men seem to be at all concerned about their fellow creatures! How completely are the lives of most men absorbed, in the pursuits of wealth, and ambition! With how many men does the love of Family, of Friend, of Country, of Mankind, appear completely impotent, when opposed to their love of Wealth, or of Power! *This is an effect of misguided association, which requires the greatest attention in Education, and Morals*" (Mill 1829: 173, emphasis added).

14. The intellectual battle "most crucial to Hogben's career was his fight against orthodox eugenics. As early as 1918 he had joined issue with those who decried the mental capacities of working class people. An additional impetus to oppose these anti-egalitarian tendencies came in 1926 when Hogben went to South Africa. There he found a new social order arising explicitly on the twin foundation of 'adventure capitalism' and scientific racism" (Werskey 1988: 105–106). "Perhaps the most important and enduring facet of Hogben's opposition to apartheid was his attempt to discredit entirely the claims of scientific racism. He carried out this attack via a relentless examination of the arguments and evidence then commonly used by eugenists" (Werskey 1988: 107). The ever-colorful Hogben apparently despised Carlyle from an early age. Though Hogben was "openly contemptuous of the 'aimless' theorizing of his colleagues, whether they were right-wing economists like Lionel Robbins and . . . Hayek, or left-wing followers of . . . Harold Laski" (Werskey 1988: 163), Peart and Levy are surely right to note that Hogben and Robbins "might have had much to say to each other. Their shared view of Carlyle would have been a beginning" (Peart and Levy 2005: 122).

15. Dickinson argues that those who "by reason of criminality, mental defect, or physical incapacity" are "socially unfit for parenthood" (Dickinson 1939: 148). He does, however, argue that the "point at issue is not the multiplication of the unfit, in the eugenic sense, but simply the responsibility for the support of children" (1939: 148).

16. Indeed, Werskey (1988: 298) notes the "anti-egalitarian views of '*scientific racists'* like Baker" (emphasis added). Mises, of course, was well aware that Carlyle's views were a forerunner of Nazi doctrine (Mises 1981: 529). Similarly, Hayek and Popper were aware of exactly what Carlylean doctrine entailed. I wonder if any discussions over eugenic planning featured in the early history of the MPS (Popper, of course, was an early member of the Society). I thank Bruce Caldwell, David M. Levy, and Sandra Peart for helpful discussion on Hayek and 1930s eugenics.

References

Carlisle, Janice. (1991). *John Stuart Mill and the Writing of Character.* Athens: University of Georgia Press.

Colander, David, Robert E. Prasch, and Falguni A. Sheth, eds. (2004). *Race, Liberalism, and Economics.* Ann Arbor: University of Michigan Press.

Cot, Annie L. (2005). "'Breed Out the Unfit and Breed In the Fit': Irving Fisher, Economics, and the Science of Heredity." *American Journal of Economics and Sociology* 64(3): 794–826.

Dickinson, H. D. (1939). *Economics of Socialism.* Oxford: Oxford University Press.

Hartwell, R. M. (1995). *A History of the Mont Pelerin Society.* Indianapolis, IN: Liberty Fund.

Hayek, F. A. (1935). *Collectivist Economic Planning.* London: George Routledge & Sons.

——. ([1940] 1997). "Socialist Calculation: The Competitive Solution." In *Socialism and War.* Vol. 10 of The Collected Works of F. A. Hayek. Ed. Bruce Caldwell. Chicago: University of Chicago Press.

——. ([1941] 1997). "Planning, Science, and Freedom." In *Socialism and War.* Vol. 10 of *The Collected Works of F. A. Hayek.* Ed. Bruce Caldwell. Chicago: University of Chicago Press.

——. ([1944] 1994). *The Road to Serfdom.* Chicago: University of Chicago Press.

——. ([1949] 1967). "The Intellectuals and Socialism." In *Studies in Philosophy, Politics, and Economics.* London: Routledge.

Kenny, Michael G. (2004). "Racial Science in Social Context: John R. Baker on Eugenics, Race, and the Public Role of the Scientist." *Isis* 95: 394–419.

Lange, O. (1936–1937). "On the Economic Theory of Socialism: Part Two." *Review of Economic Studies* 4: 123–142.

Leonard, Thomas C. (2005a). "Retrospectives: Eugenics and Economics in the Progressive Era." *Journal of Economic Perspectives* 19(4): 207–224.

——. (2005b). "Protecting Family and Race: The Progressive Case for Regulating Women's Work." *American Journal of Economics and Sociology* 64(3): 757–791.

Lerner, Abba P. (1937). "Statics and Dynamics in Socialist Economics." *Economic Journal* 47(186): 253–270.

Levy, David M. (1990). "The Bias in Centrally Planned Prices." *Public Choice* 67: 213–236.

——. (2001). *How the Dismal Science Got Its Name: Classical Economics and the Ur-Text of Racial Politics.* Ann Arbor: University of Michigan Press.

Mill, James. (1829). *Analysis of the Phenomena of the Human Mind,* vol. 2. London: Baldwin and Cradock.

——. (1992). *Political Writings.* Ed. Terence Ball. Cambridge: Cambridge University Press.

Mill, J. S. (1850). "The Negro Question." *Fraser's Magazine for Town and Country* 41: 25–31.

Mill, J. S. ([1848] 1965). *The Principles of Political Economy with Some of Their Applications to Social Philosophy.* Vols. 2 and 3 of *Collected Works of John Stuart Mill.* Ed. J. M. Robson. Toronto: University of Toronto Press.

Mill, J. S. (1986). *Newspaper Writings.* Vol. 24 of *Collected Works of John Stuart Mill.* Ed. J. M. Robson. Toronto: University of Toronto Press.

——. (1989). *Autobiography.* Ed. J. M. Robson. London: Penguin Books.

Passmore, John. (2000). *The Perfectibility of Man.* Indianapolis, IN: Liberty Fund.

Peart, Sandra J., and David M. Levy. (2005). *The "Vanity of the Philosopher": From Equality to Hierarchy in Post-Classical Economics.* Ann Arbor: University of Michigan Press.

Plato. (1987). *The Republic.* London: Penguin Classics.

Popper, Karl R. (1986). *The Open Society and Its Enemies,* vol. 1. London: Routledge.

Robbins, Lionel. (1998). *A History of Economic Thought: The LSE Lectures.* Eds. Steven G. Medema and Warren J. Samuels. Princeton, NJ: Princeton University Press.

Schumpeter, Joseph. (1946). "The Road to Serfdom by F. A. Hayek." *Journal of Political Economy* 54(3): 269–270.

Sheth, Falguni A. (2004). "John Stuart Mill on Race, Liberty, and Markets." In *Race, Liberalism, and Economics.* Eds. David Colander, Robert Prasch, and Falguni Sheth. Ann Arbor: University of Michigan Press.

Smith, Adam. ([1776] 1981). *An Inquiry into the Nature and Causes of the Wealth of Nations.* Indianapolis, IN: Liberty Fund.

Von Mises, Ludwig. (1981). *Socialism.* Indianapolis, IN: Liberty Fund.

Werskey, Gary. (1988). *The Visible College: A Collective Biography of British Scientists and Socialists of the 1930s.* London: Free Association Books.

In the Shadows of *Vanity*

Religion and the Debate Over Hierarchy

By J. DANIEL HAMMOND*

ABSTRACT. Sandra Peart and David Levy emphasize the role of economists in their excellent history of the debate between philosophers and scientists and economists and evangelicals over race and hierarchy in 19th-century Britain. Evangelical Christians have a role as allies with economists, and also Jews, but as a racial rather than a religious group. Religion, which has much to say about what it means to be human, remains in the shadows of Peart and Levy's account. The purpose of this paper is to make a start at casting light on the role of religion in the debate over race and hierarchy in 19th-century England.

> From our very infancy, on the knees of our mothers, we have been taught to believe, that to be a Catholic was to be a false, cruel, and bloody wretch; and "popery and slavery" have been wrung in our ears, till, whether we looked on the Catholics in their private or their public capacity, we have inevitably come to the conclusion, that they were every thing that was vicious and vile.
>
> William Cobbett, *A History of the Protestant Reformation in England and Ireland*

I

Introduction

RELIGION IS IN THE SHADOWS of Sandra Peart and David Levy's excellent history of the debate between philosophers and scientists and economists and evangelicals over race and hierarchy in 19th-century Britain. Peart and Levy adapt the term "philosopher" from Adam Smith's statement that "the difference between the most dissimilar characters, between a philosopher and a common street porter, for example, seems to arise not so much from nature, as from habit, custom, and

*The author thanks Don Fuey, Glenn W. Olsen, Michael Makowsky, and Anthony Waterman for comments on previous drafts.

American Journal of Economics and Sociology, Vol. 67, No. 3 (July, 2008).

education" (Smith [1776] 1976: I.2). In Peart and Levy's account, the philosophers are members of the English literary establishment, such as John Ruskin, Thomas Carlyle, Charles Dickens, and Charles Kingsley. Scientists include James Hunt, founder of the Anthropological Society of London and later professor at Columbia University, naturalist Charles Darwin, and a slew of Darwinists and eugenicists. There is considerable overlap between the Darwinists and eugenicists. Among the individuals who figure prominently are Leonard Darwin, Francis Galton, W. R. Greg, Thomas H. Huxley, and Karl Pearson. The philosophers and scientists believed that humankind is comprised of a hierarchy of races.

Classical economists who adopted Adam Smith's belief in natural equality include J. E. Cairnes, T. R. Malthus, Harriet Martineau, Nassau W. Senior, and, most prominently, J. S. Mill. These economists believed that humans come into the world as equals, and as equals share fundamental rights and capacities. The economists were joined in this belief by evangelical Christians such as William Wilberforce, Hannah More, and other so-called Exeter Hall philanthropists. The practical issues for these groups were slavery, immigration, democracy, and birth control.

Peart and Levy understandably emphasize the debate between economists and their critics. The book's subtitle is *From Equality to Hierarchy in Postclassical Economics.* It is their aim to explain why a generation of economists after the classical period went over to the side of natural heterogeneity and hierarchy. For instance, postclassical economists such as Frank A. Fetter, Irving Fisher, and A. C. Pigou enthusiastically served the cause of scientific eugenics. The evangelical Christians who were classical economists' allies in the debates over race and slavery play a supporting role to the economists' lead in Peart and Levy's history. Peart and Levy note the correspondence between the Golden Rule of Christianity and the greatest happiness principle of utilitarianism. But this is as far as religion enters the story. No religious groups other than the evangelicals play a role. Jews figure prominently, but as a racial group, not a religious group. Neither Judaism nor the Church of England nor the Roman Catholic Church merit index entries. So religion, which has much to say about what it means to be human, remains in the shadows of *The "Vanity of the Philosopher."*

The purpose of this paper is to make a start at casting light on the role of religion in the debate over race and hierarchy in 19th-century England.

II

Religion in *Vanity*

J. S. MILL WROTE in *Utilitarianism*:

> [T]he happiness which forms the utilitarian standard of what is right in conduct, is not the agent's own happiness, but that of all concerned. As between his own happiness and that of others, utilitarianism requires him to be as strictly impartial as a disinterested and benevolent spectator. In the golden rule of Jesus of Nazareth, we read the complete spirit of the ethics of utility. To do as you would be done by, and to love your neighbor as yourself, constitute the ideal perfection of utilitarian morality. (qtd. in Peart and Levy 2005: 137–138)

The equation of these two principles was a base of the anti-slavery coalition of evangelical Christians and utilitarian economists. Another base of the coalition was Adam Smith's philosophy and economics, which came into the evangelical anti-slavery movement through William Wilberforce. Wilberforce (1759–1833) was a Tory MP and member of the evangelical Anglican group known as the Clapham Sect.[1] He found in Smith's economics and philosophy ideas that seemed to complement the Calvinist doctrine of a world permeated by original sin. "To use the terms of an older Christian community, for Wilberforce Adam Smith is an inspired guide to a world without God, a doctor of the Fallen World" (Peart and Levy 2005: 167). All humans are prone to self-love, the substance of original sin. Therefore, no one is fit to be the master of others.

Peart and Levy juxtapose Robert Fellowes (1770–1847), also an Anglican, against Wilberforce and Hannah More (1745–1833). Hannah More, like Wilberforce, was an evangelical abolitionist. Fellowes was a hierarchist, but not a racist. He thought that people had different capacities for living according to the Golden Rule. Indeed, they had different capacities for accepting Christianity. Fellowes thought a certain level of civilization was a precondition for Christianity:

> For though Christianity is fitted to soften the ferocity of barbarians, if they could be brought to listen to its precepts, yet it seems better adapted and received and more likely to be practiced by those who have made some advances in civilization and in the arts of social intercourse. (qtd. in Peart and Levy 2005: 170)

Fellowes was not a Darwinist. He died 12 years before *Origin of the Species* was published. The tie between Fellowes and Darwinists was that Fellowes believed in a law of progressive human development (Peart and Levy 2005: 170). The evangelicals, on the other hand, believed that the doctrine of original sin implied that there had not been any such progressive development of human nature. All people everywhere, past, present, and future, were sinners. The famous Exeter Hall preacher Charles Haddon Spurgeon conveyed the Calvinist message of total depravity to thousands of Englishmen in sermons such as one the delivered March 7, 1858 at the Music Hall, Royal Surrey Gardens:

> Permit me to show you wherein this inability of man really does lie. It lies deep in his nature. Through the fall, and through our own sin, the nature of man has become so debased, and depraved, and corrupt, that it is impossible for him to come to Christ without the assistance of God the Holy Spirit. (Spurgeon 1858)

These are the main loci of religion in Peart and Levy's account. But there are other points of contact scattered through the book. Another of Peart and Levy's prominent subjects, Charles Kingsley, was a clergyman. Anglican clergyman, Cambridge historian and novelist, and chaplain to Queen Victoria, Kingsley appears in the book as an advocate of racial hierarchy and opponent of the economists. His story for children, *Water-Babies* (1863), was about spiritual transformation of the human body through obedience to one's betters. Other clergymen include geneticist and Catholic priest Gregor Mendel, who appears briefly in Leonard Darwin's comment that it would have been shocking for Mendel to visit his father, Charles Darwin. Presumably, Mendel would have been unsettled by Darwin's theory of evolution. Francis W. Newman, brother of Cardinal (John Henry) Newman, was one of the reformers parodied in the Cope's Tobacco literature, which Peart and Levy cover in Chapter 8, "Who Are the Canters." Newman is mentioned for having written on economics and, as a Christian, for not being a neo-Malthusian (birth control advocate). T. R. Malthus, himself an Anglican clergyman, approved of delay of marriage rather than artificial birth control for control of population. Further along chronologically in the history, Peart and Levy consider S. J. Holmes's (1939) presidential address to the American Eugenics Society, where he identifies Catholics such as G. K. Chesterton[2] as being among the few opponents of eugenics. Holmes's "list comprised: a poet, a libertarian lawyer, a Catholic literary figure, and a dead classical economist. . . . Early opposition to eugenic 'science' was remarkably thin" (Peart and Levy 2005: 118).

III

The Shadows of *Vanity*

PEART AND LEVY'S BOOK traces the analytical use of sympathy in economics from Adam Smith through the classical period, and its loss in the postclassical period, when many economists embraced eugenics. They are currently exploring the foundations of Smith's belief that sympathy is a universal component of human nature, and that, to recall Smith's example, the philosopher is fundamentally no different from and no better than the street porter (Levy and Peart 2006). Levy and Peart find the foundations of Smith's use of sympathy in Stoic philosophy. Following the Stoics, Smith elevated the ordinary experience of living one's life to the level of specialized expertise such as a philosopher might possess.

In addition to looking back in history for Adam Smith's sources, it is instructive to look to the side, to the shadows of the 18th- and 19th-century English religious landscape. During the period covered in *Vanity,* Britain was officially a Christian nation with a state church, the Church of England. Englishmen were supposed to be Christian and to live by the Golden Rule. But even so, who is one's neighbor? During at least part of the period, Britain was de facto a Christian nation as well. Owing to the political nature of the English Reformation and the national political threat perceived by many of the English from Roman Catholicism, Catholicism may have played a larger role in the debates over racial hierarchy than would be suggested by the Roman Catholic proportion of the English population and clergy.[3] Catholicism may be, in the words of that eminent Victorian, Sherlock Holmes, the dog that did not bark.

In November 1534, Parliament passed the Act of Royal Supremacy, which made Henry VIII the Supreme Head of the Church in England. Initially, the Anglican break with Rome was a matter of governance only, not doctrine. Henry had written *Assertio Septem Sacramentorum adversus Martinum Lutherum* (1521), a defense of the Catholic Church's seven sacraments against Lutheranism, for which Pope Leo X declared Henry *Fidei Defensor* (Defender of the Faith). However, as the Act gave the king authority "to visit, repress, redress, reform, order, correct, restrain, and amend all such errors, heresies, abuses, offences, contempts, and enormities, whatsoever they be," the way was cleared for doctrinal separation from Rome (Gee and Hardy 1921: 244). Three of the king's advisors and agents in the break from Rome, Thomas Cromwell, Thomas Cranmer, and Edward Seymour, were in sympathy with Protestant doctrines from the continent, and the English schism soon took on doctrinal hues. The Treasons Act, also of 1534, declared it high treason, punishable by loss of property and death, for anyone to call the king "heretic, schismatic, tyrant, infidel or usurper of the crown" (Gee and Hardy 1921: 248).

With the succession of Edward VI in 1547, the pace of doctrinal and liturgical change accelerated. Cranmer was the chief author of the *Book of Common Prayer*, whereby a communion service replaced the Latin Mass. Catholic practices, such as pilgrimages, and sacramentals, such as the blessing of ashes, were suppressed. In 1553, "Forty-Two Articles of Anglican Faith" became law under Edward. Some of the articles, such as those on the Trinity and Christ's resurrection, concerned doctrine shared by Anglicans and Catholics. Other articles bore the marks of Swiss and German Protestantism, explicitly denying Catholic doctrine and practices.[4] After Mary became Queen (1553–1558), the Act of Royal Supremacy was repealed, and England once again was officially Catholic. Elizabeth (1558–1603) restored Protestantism, and throughout her reign Anglicanism became increasingly Calvinist.

The political nature of the English Reformation is seen in the execution of Englishmen by their government in the name of the nation's religion as England shifted from Catholicism to Protestantism to Catholicism and back to Protestantism. As Gairdner (1908) observed, heresy and sedition were virtually interchangeable. From Elizabeth's reign until 1778, English Catholics had virtually no religious freedom or civil rights. Under Elizabeth, it was treasonous for an English subject to obtain Roman Catholic holy orders abroad and return to England. Giving aid to a person who did so was a capital offense. Catholic parents were forbidden from giving their children religious instruction outside their Anglican parish church. A Catholic who did not attend Anglican services was subject to a fine of 20 pounds per lunar month.[5] Under William and Mary (1689–1702), the restrictions on Catholics became less severe but more likely to be enforced. There was a 100-pound reward for anyone who aided in the arrest of a "popish" priest or bishop, and life imprisonment for any man convicted of holding holy orders in the Church of Rome. All Catholic subjects who reached the age of 18 were required to take an "Oath of Allegiance and Supremacy."

In 1778, two years after Smith's *Wealth of Nations* was published, Parliament passed the first Act for Catholic Relief. This began a gradual restoration of religious freedom and civil rights to Catholics. Among the concessions, Catholics were allowed to own and inherit land and their Protestant next-of-kin no longer had the right to enter and use their estate. A second Catholic Relief Act was passed in 1791, significantly restoring religious liberty. With the Catholic Relief Act of 1829, emancipation came close to being complete.

The Act of Union united Ireland with England in 1800, introducing to British public life a second fusion of religion and nationalism. Although Ireland gained about one-fifth of the seats in the House of Commons, the Test Act all but prohibited Catholics from occupying the seats until its repeal in 1828–1829. Soon after Dublin lawyer Daniel O'Connell was elected to Parliament in 1828, there began an Irish movement for repeal of the Act of Union and full Irish independence. Meanwhile, the potato famine in the 1840s led to massive Irish immigration to England. In 1849, the inflow of Irish immigrants was 4,000 per week (Klaus 1987: 25). The Irish population of England swelled, especially in midland cities and London. Like their English fellow countrymen, the newly arrived Irish may have been more or less devout, but their church was part of their national identity.

IV

"We'll get a rope, and hang the Pope"

THE IRISH QUESTION LOOMS LARGE in Peart and Levy's history, primarily as a racial issue, to a lesser extent as a national issue, but not as a religious issue. Were the Irish a devolved race? Were they capable of self-government? Would they respond to incentives to become more productive, or did they require direction from their betters? The Irish caricatures in *Vanity* are of a people who are primitive and brutish, impulsive and void of self-control. Peart and Levy reproduce anti-Fenian illustrations by John Tenniel from *Punch*. One shows Paddy being given the boot by John Bull for his support of Jamaicans who had been hanged. In another, Paddy is being whipped by his mistress Britannia. The apelike Irishman holds a pennant proclaiming "Fenianism For Ever." These two illustrations reveal the racial and the nationalistic side of anti-Irish sentiments. Another illustration in *Vanity* is George Cruikshank's picture of Irish bacchanalia in the palace of the Bishop of Ferns during the 1798 Rebellion[6] (from Maxwell 1845). Like the whipping of the Fenian, this picture concerns the Irish threat to British sovereignty. Peart and Levy do not remark on the revelry being set in the Catholic bishop's palace.

Racial, national, and religious matters were intertwined in English attitudes toward the Irish, French, and Italians. During the Papal State War of 1860–1861, "Punch" wrote a letter to George Bowyer, Esq., M.P. in consolation for the disloyalty of Irish troops in the Papal Brigade:[7]

> As the Pope's Knight-errant, I wonder you have not come forward to break a lance with those unworthy Irishmen—those bad Catholics—who, having apostatized from the Papal brigade and sneaked home, go about complaining of the usage which they experienced in the service of his Holiness. It seems that I must do the Holy Father's business for you. . . .
>
> They complain that some of their number were shot for breach of discipline. If so, had they not the friars of San Giovanni Decollato to confess them; and did they not therefore go to paradise? They even murmur because some of them were flogged. Their grievance was a privilege. How many holy men are continually whipping one another; how many are obliged to whip themselves, not having anyone whom they can trust to perform that pious office for them! The ungrateful grumblers ought

> to have kissed the cat-of-nine-tails which "whipped the offending Adam out of them." Excuse me for quoting a heretical poet. (*Punch*, August 4, 1860: 41)

It is reported that when Italian Risorgimento hero Giuseppe Garibaldi visited London in April 1864, thousands of children lined the streets, chanting:

> We'll get a rope,
> And hang the Pope:
> So up with Garibaldi!

A year after the tragic Irish potato famine, *Punch* published a "letter to the editor" from "Verax":

> Sir,
> In refutation of the calumny that ascribes to the Roman Catholic priesthood an influence unfavourable to the political and social advancement of a nation under its control, allow me to refer you to the history and present state of Ireland—subjects whereon the most extraordinary misconceptions prevail.
>
> Before St. Patrick converted the Irish to the holy faith of Rome, they were a set of miserable savages, but one degree elevated above the brute. Indeed, they are said to have gone actually on all fours; but I will not venture to vouch for the correctness of this last assertion; being scrupulously anxious to confine all my statements strictly to fact. . . .
>
> At the present moment we behold, in that blessed land, a spectacle of moral elevation and material greatness. Fields waving with corn attest to the perfection of agriculture. Inexhaustible peat bogs, by the aid of chemistry and capital, are converted into mines, yielding boundless wealth. . . .
>
> But, foremost among the attributes which render Ireland a model nation, is that noble sentiment which abhors and scorns deceit, prevarication, and falsehood. If there is any one trait which distinguishes the Irish character, it is scrupulous adherence to veracity. (*Punch*, vol. 20, 1851)

In the 1830s and 1840s, the Catholic presence in England grew from both Irish immigration and prominent conversions of Anglican clergy coming out of the Oxford Movement. In 1850, the Vatican restored the Catholic Hierarchy in England, making Irishman Nicholas Patrick Wiseman Cardinal and Archbishop of Westminster. The restoration with an Irishman holding the see of Westminster was referred to by Protestants as "papal aggression." Cardinal Wiseman was burned in effigy on Guy Fawkes Day, November 5, 1850. *Mr. Punch* asked what:

> His Holiness [would] say to a negro metropolitan—say a black Archbishop of Charlestown, with jurisdiction over South Carolina particularly, and in general over the whole of the Southern States of America? Make the man a Cardinal as well; give him a scarlet hat, carefully engaging him, of course, not to go hop in it. Here would be a fine opportunity of reading the Yankees a lesson of humility,—of proclaiming the great Catholic dogma of the essential equality of the human race,—and, withal, of dealing a heavy blow at slavery. Will the only answer to this suggestion be, that the idea of making a nigger a Prince of the Church, is too ridiculous? (*Punch*, vol. 19, 1850)

V

All God's Children

Mr. Punch was onto something, for the brotherhood of the entire human race was a Catholic doctrine. This principle is repeated over and over in papal encyclicals, and having been forcibly removed from the Catholic Church by the English reformers under Henry, Edward, and Elizabeth, the English people were for 300 years outside the ambit of the Catholic magisterium.

Just three years after Henry's break with Rome, Pope Paul III spoke directly to the issue that concerned 19th-century post-Darwin elitists:

> The Sublime God so loved the human race that He created man in such wise that he might participate, not only in the good that other creatures enjoy, but endowed him with capacity to attain to the inaccessible and invisible Supreme Good and behold it face to face; and since man, according to the testimony of the sacred scriptures, has been created to enjoy eternal life and happiness, which none may obtain save through faith in our Lord Jesus Christ, it is necessary that he should possess the nature and faculties enabling him to receive that faith; and that whoever is thus endowed should be capable of receiving that same faith. Nor is it credible that any one should possess so little understanding as to desire the faith and yet be destitute of the most necessary faculty to enable him to receive it. Hence Christ, who is the Truth itself, that has never failed and can never fail, said to the preachers of the faith whom He chose for that office "Go ye and teach all nations." He said all, without exception, for all are capable of receiving the doctrines of the faith.
>
> The enemy of the human race, who opposes all good deeds in order to bring men to destruction, beholding and envying this, invented a means never before heard of, by which he might hinder the preaching of God's word of Salvation to the people: he inspired his satellites who, to please him, have not hesitated to publish abroad that the Indians of the West and

> the South, and other people of whom We have recent knowledge should be treated as dumb brutes created for our service, pretending that they are incapable of receiving the Catholic Faith.
>
> We, who, though unworthy, exercise on earth the power of our Lord and seek with all our might to bring those sheep of His flock who are outside into the fold committed to our charge, consider, however, that the Indians are truly men and that they are not only capable of understanding the Catholic Faith but, according to our information, they desire exceedingly to receive it. Desiring to provide ample remedy for these evils, We define and declare by these Our letters, or by any translation thereof signed by any notary public and sealed with the seal of any ecclesiastical dignitary, to which the same credit shall be given as to the originals, that, notwithstanding whatever may have been or may be said to the contrary, the said Indians and all other people who may later be discovered by Christians, are by no means to be deprived of their liberty or the possession of their property, even though they be outside the faith of Jesus Christ; and that they may and should, freely and legitimately, enjoy their liberty and the possession of their property; nor should they be in any way enslaved; should the contrary happen, it shall be null and have no effect. (*Sublimes Dei* 1537)

Over an extended period, several popes condemned slavery: Pius II in 1462, Paul III in 1537, Urban VIII in 1639, and Benedict XIV in 1741. In 1815, Pius VII demanded that the Congress of Vienna halt the slave trade. Pius IX, who was hanged in effigy for restoring the Catholic hierarchy to England, called trade in slaves "supreme villainy" (*The Catholic Encyclopedia* 1907, "Slavery and Christianity"). His predecessor, Gregory XVI, wrote that before God, slaves and their masters are equals:

> Placed at the summit of the Apostolic power and, although lacking in merits, holding the place of Jesus Christ, the Son of God, Who, being made Man through utmost Charity, deigned to die for the Redemption of the World, We have judged that it belonged to Our pastoral solicitude to exert Ourselves to turn away the Faithful from the inhuman slave trade in Negroes and all other men. Assuredly, since there was spread abroad, first of all amongst the Christians, the light of the Gospel, these miserable people, who in such great numbers, and chiefly through the effects of wars, fell into very cruel slavery, experienced an alleviation of their lot.
>
> Inspired in fact by the Divine Spirit, the Apostles, it is true, exhorted the slaves themselves to obey their masters, according to the flesh, as though obeying Christ, and sincerely to accomplish the Will of God; but they ordered the masters to act well towards slaves, to give them what was just and equitable, and to abstain from menaces, knowing that the common

> Master both of themselves and of the slaves is in Heaven, and that with Him there is no distinction of persons.
>
> We reprove, then, by virtue of Our Apostolic Authority, all the practices abovementioned as absolutely unworthy of the Christian name. By the same Authority We prohibit and strictly forbid any Ecclesiastic or lay person from presuming to defend as permissible this traffic in Blacks under no matter what pretext or excuse, or from publishing or teaching in any manner whatsoever, in public or privately, opinions contrary to what We have set forth in this Apostolic Letter. (Gregory XVI, *In Supremo Apostolatus* 1839)

Leo XIII wrote several encyclicals bearing on human equality and slavery. In 1888, he wrote of the brotherhood of man:

> The impartiality of law and the true brotherhood of man were first asserted by Jesus Christ; and His apostles re-echoed His voice when they declared that in future there was to be neither Jew, nor Gentile, nor barbarian, nor Scythian, but all were brothers in Christ. (Leo XIII, *Libertas* 1888)

In 1890, Leo wrote to the Catholic missionaries in Africa:

> The maternal love of the Catholic Church embraces all people. As you know, venerable brother, the Church from the beginning sought to completely eliminate slavery, whose wretched yoke has oppressed many people. It is the industrious guardian of the teachings of its Founder who, by His words and those of the apostles, taught men the fraternal necessity which unites the whole world. From Him we recall that everybody has sprung from the same source, was redeemed by the same ransom, and is called to the same eternal happiness. (Leo XIII, *Catholicae Ecclesiae*, 1890)

VI

Conclusion

From the beginning of the English Reformation, Anglicanism was a mélange. Soon after Henry rebelled against papal authority, Protestant currents from Switzerland and Germany swept through England. By 1800, there were at least three "churches" within the Anglican Communion: the Low Church of Evangelical Anglicans, the High Church of Anglo-Catholics, and the Broad Church of Laditudinarians. From the first group, the evangelicals, came abolitionists such as William Wilberforce and Hannah More, and the Clapham Sect. This group tended to interpret the Bible literally. The second group included John Henry Newman, E. B. Pusey, Henry Edward Manning, and others associated with the Oxford Movement. From the High Church, there were at mid-century prominent conversions to Roman Catholicism, including Newman and Manning, both of whom became Roman Catholic cardinals. Manning was the second Archbishop of Westminster from 1865 until his death in 1892. The Broad Church was the party to which many Anglican intellectuals belonged, including John Ruskin, Thomas Carlyle, and Charles Kingsley. They tended to be rationalistic and, heavily influenced by biology and geology, regarded evangelicals' literal reading of the Bible as simple-minded.

What united these diverse groups was their Englishness. When Henry broke from Rome, declaring himself Supreme Head of the Church in England, Roman Catholicism became "foreign" and the pope became a foreign adversarial head of church and state. Moreover, the Church in England, in becoming the Church of England, was left without a magisterium. Anglican clergy and laypersons were on their own to interpret the Gospel and apply it to issues of the day. The result was twofold. When popes invoked the Catholic Church's apostolic authority to condemn slavery, the English people were unlikely to receive the news, for it was not from their church. And to the extent that information of the papal exhortation reached England, the pope's apostolic authority there was nullified by British law. If Englishmen were to conclude that slavery was wrong, or that African blacks and the Irish were their brothers, this would be on grounds other than exhortation from the Catholic Church. Not being in communion with the Church of Rome, Anglicans were without doctrinal protection from the very human temptation to treat only those humans who are like us as our brothers.

Being outside Catholic Europe, the English also lacked protection from the rationalistic idea that science and religion were incompatible. Within Anglicanism there was a divide between the evangelicals, who placed their faith in the Bible, and the Broad Churchmen, who were willing to mold their faith to the dictates of science. For Catholics, on the other hand, there was no incompatibility between faith and science, or more broadly between faith and reason. Pope Leo XIII wrote in his encyclical *Providentissimus Deus* (1893):

> There can never, indeed, be any real discrepancy between the theologian and the physicist, as long as each confines himself within his own lines, and both are careful, as St. Augustine warns us, "not to make rash assertions, or to assert what is not known as known."

Leo quoted St. Augustine's rule for theologians regarding natural science:

> Whatever they can really demonstrate to be true of physical nature, we must show to be capable of reconciliation with our Scriptures; and whatever they assert in their treatises which is contrary to these Scriptures of ours, that is to Catholic faith, we must either prove it as well as we can to be entirely false, or at all events we must, without the smallest hesitation, believe it to be so.

Commenting on English "theophobia," Catholic convert and polymath, Sir Bertram Windle, wrote: "To set up science as an 'unknown God' seems a curious choice, even more curious than the choice of humanity, which—pitiable object as it is—was at least made in the image of God" (1919: 49).

Notes

1. Among other members of the Clapham Sect was the banker/economist Henry Thornton.

2. See Chesterton (1922).

3. In 1850, there were 788 Catholic clergy in England and Wales, with 168 in the London district. In an 1851 religious census, Catholics made up roughly 3.5 percent of all persons attending church (Norman 1984: 205–206).

4. *Sola fide* (articles 11, 12, and 13); errors in matters of faith by the Church of Rome (article 20); denial of the doctrine of purgatory (article 23); prohibition of the use of Latin (article 25); denial of five of the seven sacraments (article 26); the claim that the doctrine of transubstantiation "is repugnaunt to the plaine woordes of Scripture" (article 29); and renunciation of clerical celibacy (article 31).

5. There being 13 lunar months in a year.

6. Catholic Diocese of Ferns, province of Leinster.

7. M.P. for Dundalk, County Louth. Bowyer was an English baronet, lawyer, and, in his youth, member of the Oxford Movement. He converted to Catholicism in August 1850 and served as constitutional advisor to Cardinal Manning.

References

The Catholic Encyclopedia. (1907). Rpt. in *New Advent* (2007) http://www.newadvent.org/cathen/index.html.

Chesterton, G. K. ([1922] 1986). "Eugenics and Other Evils." In *The Collected Works of G. K. Chesterton,* vol. 4. San Francisco: Ignatius Press.

Cobbett, William. ([1824] 1998). *A History of the Protestant Reformation in England and Ireland.* London: Pickering & Chatto.

Gairdner, James. (1908). *Lollardy and the Reformation in England: An Historical Survey.* London: Macmillan.

Gee, Henry, and William John Hardy. (1921). *Documents Illustrative of English Church History.* London: Macmillan.

Holmes, S. J. (1939). "The Opposition to Eugenics." *Science* 89: 351–357.

Kingsley, Charles. (1863). *The Water-Babies: A Fairy Tale for a Land-Baby.* London: Macmillan.

Klaus, Robert J. (1987). *The Pope, the Protestants, and the Irish: Papal Aggression and Anti-Catholicism in Mid-Nineteenth Century England.* New York: Garland.

Levy, David M., and Sandra J. Peart. (2006). "Adam Smith & His Sources: The Evil of Independence." Paper presented at the Summer Institute for the Preservation of the History of Economics, George Mason University, June 20.

Maxwell, W. H. (1845). *History of the Irish Rebellion in 1798: With Memoirs of the Union, and Emmett's Insurrection in 1803.* London: Baily Brothers.

Mill, John Stuart. ([1861] 1969). "Utilitarianism." In *Collected Works of John Stuart Mill,* vol. 5. Ed. John M. Robson. Toronto: University of Toronto Press.

Norman, Edward. (1984). *The English Catholic Church in the Nineteenth Century.* Oxford: Clarendon Press.

Papal Encyclicals Online. (2007). http://www.papalencyclicals.net.

Peart, Sandra J., and David M. Levy. (2005). *The "Vanity of the Philosopher": From Equality to Hierarchy in Postclassical Economics.* Ann Arbor: University of Michigan Press.

"Popery and Progress." (1851). *Punch.*

"The Pope's Irish Ragamuffins." (1860). *Punch* August 4: 41.

Smith, Adam. ([1776] 1976). *An Inquiry into the Nature and Causes of the Wealth of Nations.* Ed. W. B. Todd. Oxford: Clarendon Press.

Spurgeon, C. H. (1858). "Human Inability." Sermon 182. http://www.spurgeon.org/sermons/0182.htm.

Windle, Sir Bertram C. A. (1919). *Science and Morals and Other Essays.* London: Burns & Oates.

The Vanity of the Economist

A Comment on Peart and Levy's The "Vanity of the Philosopher"*

By KEVIN D. HOOVER*

ABSTRACT. In the *Vanity of the Philosopher*, Sandra Peart and David Levy reconsider "postclassical" economics from the vantage point of Adam Smith's "analytical" egalitarianism. Analytical egalitarianism is assumed, not proved; and Peart and Levy's criticisms of many 19th- and early 20th-century economists, as well as eugenics in general, depend on equivocating between analytical and substantive egalitarianism. They fail to provide a non–question-begging critique of eugenics.

Peart and Levy's (2005) *The "Vanity of the Philosopher"* is a wonderfully rich tapestry, full of historical detail and intellectual insight. But it is also a work of persuasion, aiming to convince the reader of a larger story. That story, unhappily, is not fully fleshed out, the plot not fully convincing, and the moral not compelling. I reach these conclusions as a friendly critic who is deeply sympathetic to many of their attitudes and substantive beliefs.

Levy and Peart define *analytical egalitarianism* as the doctrine that takes as its working assumption that people are to be regarded as homogeneous in capabilities and respect. To assume that people are arranged into moral, intellectual, or social hierarchies is the opposite of analytical egalitarianism. *Vanity* explores the role of the competing assumptions of analytical egalitarianism and hierarchy in the debates over race (Irish versus English, African versus European) among 19th-century British economists and public intellectuals and the role of economists in the late 19th- and early 20th-century eugenics movement.

*The author is Professor of Economics and Philosophy at Duke University. This paper is based on comments given in the Roundtable on the Vanity of the Philosopher at the History of Economics Society Annual Conference at Grinnell College, June 23–26, 2006.

American Journal of Economics and Sociology, Vol. 67, No. 3 (July, 2008).

Peart and Levy divide economics into "classical" and "postclassical"—roughly into before and after John Stuart Mill. Classical economics was, they argue, grounded in analytical egalitarianism; postclassical, in the assumption of hierarchy.

Adam Smith and John Stuart Mill are the heroes of *Vanity.* Peart and Levy take Smith's ([1776] 1976: Bk. 1, Ch. 2) formulation of egalitarianism as their starting point and as the source of their title:

> The difference between the most dissimilar characters, between a philosopher and a common street porter, for example, seems to arise not so much from nature as from habit, custom, and education. When they came into the world, and for the first six or eight years of their existence, they were perhaps very much alike, and neither their parents nor playfellows could perceive any remarkable difference. About that age, or soon after, they come to be employed in very different occupations. The difference of talents comes then to be taken notice of, and widens by degrees, till at last the vanity of the philosopher is willing to acknowledge scarce any resemblance.

Note, in contrast to Peart and Levy's implication, that Smith is not here advocating analytical egalitarianism. Instead, Smith acknowledges that people are substantively unequal and analyzes the source of that inequality, concluding that it is nurture, not nature. The equality that he identifies is an equality of the original potential of people—a substantive, not an analytical, egalitarianism. Equivocation between analytical and substantive egalitarianism does much of the work in *Vanity.*

Peart and Levy's other hero, Mill, played a similar role in Levy's earlier *How the Dismal Science Got Its Name* (2001). Mill is the man of action, having fought the good fight against slavery and, in the case of John Eyre, the governor of Jamaica on trial for a massacre, on the side of equal justice for former slaves.

The villains are Thomas Carlyle and John Ruskin—again, principals in Levy's *Dismal Science*—and Adolph Hitler. The case against Carlyle and Ruskin as racists is explicit. The case against Hitler is implicit in the question: "Does the Holocaust provide a firewall to eugenics?" (Peart and Levy 2005: 125, fn. 32; cf. 110). Peart and Levy write: "Firewalls do not maintain themselves. One purpose of our book is to help maintain a firewall in the space of ideas by discussing the consequences that have followed from the assumption that surface differences among people reveal underlying differences among persons" (2005: 125). Clearly, *Vanity* is not a dispassionate work. Peart and Levy do not hide their passionate opposition to racism and eugenics; nor do they disguise their polemical tone. There would be no objection to such a style, except where it distorts vision. Unfortunately, I think that, in Peart and Levy's case, it does just that.

The argument of *Vanity* stripped to its bones runs like this: Adam Smith was a proponent of analytical egalitarianism; and, for Smith, sympathy is the fundamental moral category (Smith 1759). Thomas Carlyle is no kind of egalitarian, does not approach "inferior" races sympathetically, and—as we can see perfectly clearly in the 21st century—was on what we now regard as the wrong side of the racial politics of the 19th century, having supported slavery and an inferior legal and social status for Africans, the Irish, and other "inferior" races. People who thought rather like Carlyle on the matter of social hierarchy were responsible for the transformation of classical into postclassical economics *and* they supported eugenics. Hitler infamously supported eugenics, with the most dire moral consequences. Since slavery and Hitler are evil, postclassical economics is tainted.

Leaving aside a great deal of interesting details, that *is* the argument—not, I think, an unfair caricature. Put baldly, it is clearly a weak argument. When my daughter tries to convert me to vegetarianism, I remind her that Hitler was a vegetarian. But naturally, I jest: She understands that my argument is not a morally serious one. Unhappily, Peart and Levy's argument takes a similar form.

The fundamental problem is that Peart and Levy never engage the central issue by providing a non–question-begging account of exactly what is wrong with eugenics per se. They get a lot of mileage out of associating eugenics with Hitler, but most eugenicists were not mass murderers or advocates of mass murder; and, while it is true that Hitler gave a eugenic justification for his policies, Peart and Levy do not establish that a holocaust is the inevitable endpoint of every eugenic doctrine.

Indeed, Peart and Levy simply "plunk" for Smith's vision of the human moral condition without examining it in any detail, much less providing a compelling argument for its truth. That they are aware of the softness of these foundations is suggested by their decision to call their preferred position "analytical" egalitarianism. The qualifying adjective serves to remind us that egalitarianism is offered as a methodological strategy rather than a substantive conclusion. The injunction to construct economic analyses as if people were homogeneous is quite different from asserting that they really are homogeneous.

Smith conjectured that the real differences between people were grounded in nurture, not nature, but he did not suggest that they were not real after all. Frankly, it is prima facie absurd to assume that there are no real differences. The questions have to be: What is the role of habit, custom, and education in comparison to biology in supporting real differences? And are any substantial biological differences stably associated with race, as defined by socially significant visible or national differences? Postclassical economists are tarred with support for sterilization, race-based immigration, and prejudice in favor of eugenic results, yet their substantive arguments with regard to the underlying question of egalitarianism versus hierarchy are not on the table.

But this is an objectionable mode of argument. First, Peart and Levy really need to address the substance of the arguments rather than the motives of the postclassical economists, whose "vanity" is to advocate a hierarchy in which they stand rather high themselves. The philosopher Sidney Hook put it well in his "Ethics of Controversy" (1954): "*Before* impugning an opponent's motives, even when they legitimately may be impugned, answer his arguments."

Second, Peart and Levy implicitly connect postclassical economics to eugenics and eugenics to bad outcomes (such as the Holocaust). This is not persuasive because, on the one hand, they never really clarify exactly what postclassical economics comprises and, on the other hand, they do not establish a deep connection between its doctrines and eugenics. Is modern microeconomics (as represented, for example, in a textbook such as Varian 1984) implicated in their argument? I doubt that Peart and Levy think so. Yet, we think of Edgeworth and Marshall, for instance, as direct predecessors to modern microeconomics. If modern microeconomists are not closet eugenicists, it suggests that postclassical economics *qua* economics may not, as Peart and Levy imply, have been as organically related to whatever eugenic notions that its creators may also have held.

I should declare a personal interest. I am a child of eugenics. I am the fifth of eight children. My mother always told us that one reason that our family was large was that, as a sociology major in the early 1940s, her teachers had told her that educated people were not having enough children. This story received some confirmation when, at the time of my parents' 50th wedding anniversary, my older sister produced a *Look Magazine* with a popular article by the anthropologist Margaret Meade (1944), making exactly that case for *Look*'s—presumably educated—readers.[1]

My story is relevant as it points out that eugenics is still with us and that its supports may be popular as well as hierarchical. For example, abortion or genetic engineering for preferred traits or against "defect" are already widely practiced—and not under pressure from the government, but from the grassroots. A recent paper (Ananat et al. 2006) confirms the findings of Donohue and Levitt (2001) that the legalization of abortion reduced crime rates by prospectively reducing a segment of the population likely to engage in criminal behavior. The authors also provide evidence that legal abortion raises college graduation rates, lowers welfare use, and lowers the number of children in single-parent households.[2] While eugenics is not mentioned—even *sotto voce*—the analysis is in precisely the same spirit as Dugdale's calculations of the social costs of the "Juke" family and of the social benefits from sterilizing them (cited by Fisher 1909; see Peart and Levy 2005: 115).

Eugenics has been fostered by governments and experts; but, in an age of increasingly available information about individual genetics, it is frequently homegrown and popular. Peart and Levy loathe eugenics and put their faith in egalitarianism. But what happens if that faith is misplaced, and democracy and individual members of society abandon them?

The difficulty seems to me to be that Peart and Levy try to make the case against racism and eugenics on the procedural basis of analytical egalitarianism. Yet, what they want to conclude is that a *substantive* egalitarianism is more or less true. They have not really made the case for substantive egalitarianism from their "analytical" premise, and they equivocate: While they ostensibly appeal to analytical egalitarianism, they nonetheless assume that substantive egalitarianism is established, and damn eugenics without further discussion. In a passage quoted earlier, Peart and Levy draw the distinction between "surface differences among people" and the "underlying differences among persons." The surface differences among people are palpable and, historically, socially freighted. The important issues, however, concern the reality of the underlying differences between people and, if they are real, how they might be connected to the surface differences. The only sensible course for anyone concerned about the recrudescence of eugenics is to engage in a serious investigation of the truth of eugenics and in a serious debate over substantive egalitarianism.

Since my own sympathies are strongly anti-eugenic, recent trends are discouraging. Still, Peart and Levy's approach does not buck me up: Eugenic issues are upon us—firewalls or no firewalls—whether we like it or not, but scientific discussion is almost impossible. For example:

- The *Wall Street Journal* (Regalado 2006) recently reported that Bruce Lahn, an evolutionary biologist/geneticist who published evidence on racial divergence in recent human evolution, has turned to completely different lines of research because of "politically correct" attacks on his work. Apparently, some plausible scientific hypotheses cannot be investigated.
- Economics is not immune. A recent exchange on the History of Economics listserv failed to engage the substance of the evidence on substantive racial differences and strayed into pure assertion, appeals to authority, and ad hominem attacks on those who considered that racial differences might possibly be genuine (http://www.eh.net/pipermail/hes/; initial entry by Himoe 6 February 2006; followed up by entries under various subject lines by Dimand, Forget, Peart, and Shah, all in February). Here is certainly a case in which Hook's dictum ought to apply.
- The intellectual mood is well captured in a recent open letter from the Dutch ambassador to the United States, in which he defended his government's behavior with respect to the recent emigration (exile?) of Hirsi Ali, a member of the Dutch parliament, progressive Muslim woman, and critic of Islamic

> conservatives. The ambassador's letter concluded with a clarion: "intolerance will not be tolerated" (Eenennaam 2006). It is one thing not to tolerate death threats, but quite another to shut down inquiry because one does not like what might be found.

It does not help to shout them down. Let us pursue the truth and shame the devil.

It is possible to be in a very similar place to Peart and Levy politically—as I imagine that I am—and yet to be unconvinced by the argument that takes substantive egalitarianism (as well as the substantive falseness of eugenics) as its hidden premise while declaring that it is grounded in analytical egalitarianism. If substantive egalitarianism is true (and eugenics false), then it needs to be demonstrated by argument and evidence. Nevertheless, many of the political consequences that Peart and Levy look for are more solidly grounded in simple humility. Peart and Levy quote Lionel Robbins (1938: 635; Peart and Levy 2005: 209): "I do not believe, and I never have believed, that in fact men are necessarily equal or should always be judged as such. But I do believe that, in most cases, political calculations which do not treat them as if they were equal are morally revolting." Though Robbins is cited favorably, his argument is not Peart and Levy's. It does not presume that there are no racial differences or that eugenics is substantively wrong. It is rather an argument consistent with the view that on vexed questions there is some truth of the matter, and we should seek the truth, yet we should not presume that we have got it or legislate on the basis that we could not be mistaken.

Smith originally conjectured that differences among people were not inbred, which is an argument for substantive equality among races, if not, ex post, among individuals. The abolitionist cry, "Am I not a man and a brother" (2005: 188, 189) appeals to such substantive equality, and is the sort of egalitarianism that does the work for Peart and Levy. In his famous *Essay on the Nature and Significance of Economic Science* (1935) as well as in the comment cited by Peart and Levy (Robbins 1938), Robbins argues for the inscrutability of personal utility (cf. Peart and Levy 2005: 209), driving a familiar wedge between scientific, positive economics and normative policy. For Robbins, unlike Smith, for whom sympathy was a fact of human nature, egalitarianism could be only analytical. The supposition of analytical egalitarianism could provide no argument against eugenics or inherent differences in racial capacities, but neither could the truth of biological difference in itself provide an argument for racism or social hierarchy.

Notes

1. Meade (1944) does not make an explicitly eugenic argument, and the bulk of the essay makes the case for the welfare state, for wide governmental support for childrearing. Nonetheless, she is explicit that the ratio of rich and middle-class to lower-class birthrates is too low and that it is important to raise it. Were nurture everything, then she would have no ground to insist on more children in the higher ranks, provided that adequate public resources were made available to poor families. Yet she envisages public resources going to the middle classes as well as the poor, with the aim of shifting the balance toward the higher social classes, which is just a softer version of suppressing population growth among the poor.

2. Eastland (2004) argues that another effect of legal abortion is to reduce the number of Democratic voters: There is a high correlation between the voting patterns of parents and children, and Democrats have had higher rates of abortion than Republicans after *Roe v. Wade*. Brooks (2006) points out that liberals have a lower fertility rate than conservatives.

References

Ananat, Elizabeth Oltmans, Jonathan Gruber, Philip B. Levine, and Douglas Staiger. (2006). "Abortion and Selection." National Bureau of Economic Research, Working Paper No. W12150, April.

Brooks, Arthur C. (2006). "The Fertility Gap: Liberal Politics Will Prove Fruitless as Long as Liberals Refuse to Multiply." *Wall Street Journal: Opinion Journal* 22 August; available at http://www.opinionjournal.com/editorial/feature.html?id=110008831.

Donohue, John J., III, and Steven D. Levitt. (2001). "The Impact of Abortion on Crime." *Quarterly Journal of Economics* 116(2): 379–420.

Eastland, Larry L. (2004). "The *Roe* Effect: The Empty Cradle Will Rock: How Abortion is Costing the Democrats Voters—Literally." *Wall Street Journal: Opinion Journal* 22 August; available at http://www.opinionjournal.com/extra/?id=110005277

Eenennaam, Boudewijn J. van. (2006) "Statement by Boudewijn J. van Eenennaam, Ambassador of the Kingdom of the Netherlands to the United States, May 19, 2006." Royal Netherlands Embassy, Washington, D.C.

Fisher, Irving. (1909). *National Vitality: Its Wastes and Conservation.* New York: Arno.

Hook, Sidney. (1954). "The Ethics of Controversy." Rpt. in Hook (1980) *Philosophy and Public Policy*. Carbondale, IL: Southern Illinois University Press.

Levy, David M. (2001). *How the Dismal Science Got Its Name: Classical Economics and the Ur-Text of Racial Politics.* Ann Arbor: University of Michigan Press.

Meade, Margaret. (1944). "Can American Afford Families? We Can't Afford Not to Have Larger Middle Class Families." *Look Magazine* 12 December: 90.

Peart, Sandra J., and David M. Levy. (2005). *The "Vanity of the Philosopher": From Equality to Hierarchy in Post-Classical Economics.* Ann Arbor: University of Michigan Press.

Regalado, Bruce. (2006). "Head Examined: Scientist's Study of Brain Genes Sparks a Backlash: Dr. Lahn Connects Evolution in Some Groups to IQ; Debate on Race and DNA: 'Speculating is Dangerous'." *Wall Street Journal* 16 June: A1+.

Robbins, Lionel. (1938). "Interpersonal Comparisons of Utility: A Comment." *Economic Journal* 48(192): 635–641.

Smith, Adam. (1759). *The Theory of Moral Sentiments.*

——. (1776). *An Inquiry into the Nature and Causes of the Wealth of Nations.*

Varian, Hal R. (1984). *Microeconomic Analysis.* New York: Norton.

Classical Equality

On the Content of Analytical Egalitarianism

By JOSEPH PERSKY*

ABSTRACT. Sandra Peart and David Levy in *The "Vanity of the Philosopher"* champion a concept of "analytical egalitarianism." Equality is a difficult concept. Peart and Levy attempt to reconstruct analytical egalitarianism from the classical writing of British political economy from Adam Smith to John Stuart Mill. Aspects of this reconstruction touch on a number of different egalitarian conceptions, including: (1) equality of capacity and talent, (2) racial equality, (3) equality in the marketplace, (4) equality of opportunity, (5) equality of material conditions, (6) equality of happiness, (7) equality before God, and (8) political equality. This paper briefly considers the relation of each of these equalities to Peart and Levy's analytical egalitarianism. The hope is that such exercises can help elucidate Peart and Levy's reinterpretation of classical economic's understanding of equality. A central theme does emerge. Peart and Levy, echoing the classical economists themselves, seem reluctant to follow their radical assumptions concerning talent and capacity for happiness to radical conclusions concerning the appropriate provenance of redistributional policies.

I

Introduction

AMONG PHILOSOPHERS, it has become something of a commonplace that "equality" is a particularly difficult word.[1] Equality carries a range of meanings and connotations. Still, how we conceptualize equality does matter, precisely because an agreed definition seems to require a political commitment.[2] Sandra Peart and David Levy in *The "Vanity of the Philosopher": From Equality to Hierarchy in Post-Classical Economics* champion a concept of "analytical egalitarianism." Analytical

*Joseph Persky teaches in the Department of Economics, University of Illinois at Chicago.

American Journal of Economics and Sociology, Vol. 67, No. 3 (July, 2008).

egalitarianism is reconstructed from the classical writing of British political economy from Adam Smith to John Stuart Mill. As such, it doesn't correspond exactly to any of the crop of currently popular "equalities." Nevertheless, given its breadth, aspects of this reconstruction touch on a number of different egalitarian conceptions, including: (1) equality of capacity and talent, (2) racial equality, (3) equality in the marketplace, (4) equality of opportunity, (5) equality of material conditions, (6) equality of happiness, (7) equality before God, and (8) political equality.

This paper briefly considers the relation of each of these equalities to Peart and Levy's analytical egalitarianism. The hope is that such exercises can help elucidate Peart and Levy's reinterpretation of classical economic ideas about equality. Amid these peregrinations, a central theme does emerge. Peart and Levy, echoing the classical economists themselves, seem reluctant to follow their radical openings to logical conclusions.

II

Equalities

A. *Equality of Capacity and Talent*

If *Vanity* did nothing else, it would be a valuable contribution for its drawing sharp and long overdue attention to Adam Smith's radical position on the equality of natural talents.[3] In the Smith quote that introduces Chapter 1 of *Vanity* and contributes a central phrase ("the vanity of the philosopher") to the book's title, Smith compares the street porter to the philosopher and concludes that before education they were much the same. The passage (according to Edwin Cannan in his famous edition of Smith's *Wealth of Nations* (1776)) was probably written in opposition to Joseph Harris's *Essay Upon Money and Coins* (1757).

Harris anticipated Smith in emphasizing the importance of the division of labor to modern economies. But unlike Smith, Harris rooted the division of labor in people's intrinsic differences: "Men are endued [*sic*] with various talents and propensities, which naturally dispose and fit them for different occupations" (1757: 15). The social process of specialization sets the foundation for the accomplishments of true genius. Harris cites several accomplished mechanics. He is rapturous over the achievements of Isaac Newton. These men were "great benefactors to their country" to which they brought "great reputation and wealth" (1757: 16).

In fairness to Harris, he doesn't flail the point, but goes on to add a number of practical considerations having little to do with intrinsic ability. Chief among these, he recognizes the implausible difficulty of one person learning all the skills necessary for their own support. In these comments he sounds much like Smith (or given Harris's earlier date of publication, Smith sounds much like Harris). Harris considers these arguments cumulative and mutually supportive. He draws no sharp opposition between nature and nurture.

But Smith draws just such a contrast. He makes a loud argument for the role of education and other aspects of nurture in shaping our skills and talents. In this observation (again, according to Cannan) he can be recognized as "in accordance with the view of Hume."[4] Hume's comments on "how nearly equal all men are in their bodily force, and even in their mental powers and faculties, till cultivated by education" (Hume 1752: II, xii, 4) arose in a discussion of possible initial social contracts, not the division of labor. Yet the thought very much resonates with Smith. Indeed, it resonates so completely that one wonders if Smith hasn't gone a good deal further than he meant. Perhaps in making a point about the division of labor, he has overstated his case and suggested more than he fully believed.

I suspect that Smith did say more here than he was prepared to defend. He would not be the first to exaggerate his position to make a clean point. The tension is suggested by Smith's own discussion of wage differentials later in the *Wealth of Nations*. Smith begins, consistently enough, by arguing the role of compensating differentials. In a society of "perfect liberty," he argues, "[i]f in the same neighborhood, there was any employment evidently either more or less advantageous than the rest, so many people would crowd into it in the one case and so many would desert it in the other, that its advantages would soon return to the level of other employments" ([1776] 1937: 138). And he starts very much in this vein with what are clearly compensating differentials: the unpleasantness of an activity, the costs of learning a profession, and the constancy of employment. But then Smith goes on to consider also the "probability of success." Smith puts forward an argument that risky professions and activities are not fully compensated because people overestimate their chances of success. The returns are high, but not as high as might otherwise be expected. For the most part, Smith seems to suggest success in these professions is largely random. But then he waffles: "To excel in any profession, in which but few arrive at mediocrity, is the most decisive mark of what is called genius or superior talents." This seems more than education. And when he moves on to "players, opera singers, opera dancers, etc," the notion of some sort of natural talent is bubbling up through the text. Smith comments on the "rarity and beauty of the talents" and seems almost ready to go over to Harris's position. But then he pulls back a bit: "Such talents, though far from being common, are by no means as rare as is imagined. Many people possess them in great perfection, who disdain to make this use of them; and many more are capable of acquiring them, if any thing could be made honourably by them" ([1776] 1937: 149). Many could acquire these talents, but not all.

Smith draws a picture of fairly tight talent distributions, with some outliers. Smith's position is optimistic, but not doctrinaire. We are far more similar in native ability than we think. There is a strong optimism in this position, an optimism that similarly infuses Peart and Levy's analytical egalitarianism. But at the same time, Smith's position is quite radical. From a utilitarian perspective, it sets the stage for considerable disappointment if the economic institutions of society fail to appreciate the underlying similarities among people. I'm not trying to argue that Smith accepted such radical conclusions;[5] indeed, he tried mightily to avoid them by placing the fault not on the market itself, but on the malfeasance of a mercantilist elite.[6] Whatever the plausibility or implausibility of that claim, there is a tension here in Smith, one shared by many of his classical followers and most notably John Stuart Mill. This same tension suffuses Peart and Levy.

B. Racial Equality

Peart and Levy come to this volume with a deep appreciation of mid-19th-century British racism and the nobility of the liberal opposition to such bigotry. (See in particular Levy's *How the Dismal Science Got Its Name: Classical Economics and the Ur-Text of Racial Politics*, 2001.) They know this material. A good portion of the *Vanity of the Philosopher* aims at extending this history as racism drew strength from late 19th-century statistics, anthropology, sociology, and evolutionary biology. Peart and Levy plausibly identify two forms of racism (2005: Ch. 4). Under one view, lesser races are simply blocked in their development. Under the other, the "mean" is shifted "down," but considerable variability exists within each race. Peart and Levy do an excellent job in unearthing and explicating the debate between these camps.

One possible interpretation of analytical egalitarianism is that it holds up in opposition to both these racisms a concept that suggests that most large groupings of people will be distributed pretty much the same across a wide range of characteristics and potentials. This might be called sociological equality. Most of us are comfortable with such a worldview. For many it captures the essence of the liberal position. At several points in *Vanity*, Peart and Levy seem to endorse this sociological interpretation of equality. The sociological view doesn't assert equality across individuals, but only as between groups. Does this do justice to Smith's position? Does this appropriately capture the import of classical analytical egalitarianism?

Somewhat surprisingly, the discussion of this type of sociological equality seems to originate with the sociological school of the southern United States, and more specifically in the 19th-century work of George Fitzhugh (1854). In discussing apparent differences between the Northern and Southern states before the Civil War, Fitzhugh, one of the country's first sociologists, argued that institutional differences must have accounted for the economic success of the former, since each region (or at least the whites in each region) could be expected to show much the same distribution of abilities. Fitzhugh went on to suggest that many white households, in North and South, were incompetent and could benefit from the institution of slavery.

This isn't meant to tar the sociological conception of equality with Fitzhugh's proto Dorian ethic. But hopefully it does drive home the point that equality of distributions still leave open the question of how large are the tails of that common distribution and, if substantial, how best are we to deal with those tails. The sociological view may be color-blind, but it hardly seems like equality at all. It certainly falls short of Smith's claim that, for the most part, we are pretty much the same.

C. Equality in the Marketplace

At times it seems that Peart and Levy seem to define analytical egalitarianism as little more than a commitment to equality in the marketplace, a classical anticipation of Milton Friedman's freedom to choose: "we hold that the classical economists got it right: an analytical system in which everyone counts equally and is presumed equally capable of making decisions is the only system that seems morally defensible to us" (Peart and Levy 2005: xiv). If each of us is the best judge of his or her own welfare, a general freedom to choose one's own way promises considerable utilitarian value. J. S. Mill's economic men, competent wealth-maximizers subject to interests in leisure, luxury, and sex, need room to exercise their decision-making faculties. For them, equality extends only to choice. Viewed from a utilitarian perspective, it holds that the freedom of the marketplace is the critical ingredient in achieving the greatest good of the greatest number of a large group of differently situated individuals competent to judge their own interests. People are not the same. They are different in important ways, yet equal in their ability to discern their own differing interests in a competitive market setting. This is a common interpretation of 19th-century liberalism. It vouches the poor man the same right to contract as the rich man.

Equality in the market seems to turn equality upside down. Admittedly, classical economists did often argue that the freedom to contract was of fundamental importance to the welfare of the poor. For example, Jeremy Bentham in his *Defense of Usury* (1787) looked on any infringement on the poor's right to contract as a breach of their basic liberties. But can this be the meaning of equality: the right to make the best of a small lot in the marketplace? If so, equality is little more than a cruel joke. And if Smith himself sometimes seemed to argue just this limited sense of equality, why did he need such a radical statement concerning the underlying similarities between the street porter and the philosopher? Only to grant the street porter a freedom in the marketplace?

Certainly, his radical endorsement of equality of intrinsic talents isn't required to yield such small pottage. Perhaps Peart and Levy have just followed Smith into rhetorical overkill, but it seemed like they (and he) were stalking much bigger game.

D. Equality of Opportunity

If equality in the marketplace strikes us as a somewhat pale prescription for inequality, equality of opportunity has a distinctly more promising ring. Smith himself and the majority of his classical followers strongly endorsed the value and rightness of public education (see Miller 1966). There is a clear appreciation (now well acknowledged) in both Smith and J. S. Mill of the mind-deadening quality of much modern work. Mill is particularly eloquent as to the reasons for parents' underinvestment in their children's education. To make market equality meaningful, extensive investments in public education are mandatory. This story is there in the classical writers. But, as committed as they were to such public expenditures, they hesitate to put these thoughts at center stage. Indeed, their treatment of public education has the flavor of an afterthought.

If public education was to be the basic means for achieving more equal outcomes consistent with Smithian notions concerning the equality of talent, the argument would seem to require a stronger presentation than Smith or Mill provided. The classical economists seem almost apologetic that equality of markets proves unable to do the job by itself. Education and a few minor welfare reforms are added to address the deficiency, but only in a rather half-hearted fashion. Nor is there any serious consideration of the extent to which access to formal education must be matched to a certain basic level of material welfare. We can only wonder why such steps play such a limited role in the realization of classical analytical egalitarianism.

E. Equality of Material Conditions

Equality of opportunity is handled jauntily by the classical writers. Efforts to more immediately establish material equality are attacked outright. Despite their strongly expressed preference for material equality, the classical economists immediately deny the possibility or desirability of trying to achieve such equality in the short run. Indeed, it is not too much to say they were appalled by such proposals. Their fundamental concerns seem to have been that (a) redistribution might lead to a rather speedy return to inequality with little being accomplished overall, (b) redistribution might lead to an erosion of private incentives resulting in some sort of stagnation, or, worst of all, (c) radical redistribution might result in the creation of an anarchic state, a return to an all-against-all as the social structure breaks down (see Persky 2006). Peart and Levy have little to say about the degree of direct material redistribution implied by their "analytic egalitarianism." Presumably, it's not much. But if this is the case, it seems incumbent on them to lay out the classical argument against such redistribution and to comment on the cogency of those arguments, given the presumed equality of talent.

F. Equality of Happiness

This concept and the closely related equality of the capacity for happiness that Peart and Levy address draw heavily on Smith's commitment to a sympathetic psychology. Presumably, sympathy alerts us to the general capacity of others for happiness. Hence, analytical egalitarianism involves an assumption of equal capacity.[7] However, the discussion in Peart and Levy doesn't seriously touch on the question of equal outcomes. If we grant a rough equality of capacity for happiness, shouldn't utilitarians accept a social responsibility to achieve something close to an equality in the actual distribution of happiness?

This question, at first, seems only a restatement of the tension between equality in talent and a social responsibility for equality in material outcomes. Or, even worse, it might involve us in debates over whether society is under a responsibility to subsidize those with highly refined and expensive tastes, people unable to achieve average levels of happiness without consuming fine wines and diamond tiaras. More seriously, equality of happiness raises questions about the social obligation to offset handicaps and disabilities. Such matters form a key portion of Dworkin's argument against a happiness standard for equality (Dworkin 2000: Ch. 1). Equality of happiness may prove a much more difficult goal than equality of material conditions.

But there is a very different utilitarian tradition on the equality of happiness that emphasizes not the difficulty of achieving such a standard, but the ease. Among the earliest utilitarians, the political philosopher Claud-Adrien Helvetius in the mid-18th century held that achieving equality of material outcomes was simply infeasible. The drive to acquire wealth catches up many people; some achieve wealth, most don't. Perhaps not surprisingly, wealth holders resist powerfully the full redistribution of their spoils. Anticipating later utilitarian arguments by Bentham and J. S. Mill, Helvetius held that efforts to level material differences could only produce discord and worse. However, material equality, according to Helvetius, was not necessary for something approaching equality of happiness. The idle rich were often unhappy, subject to a boring ennui. For most of us, a reasonable amount of work provides a sense of achievement and meaning to the average life. A pleasant level of happiness can be achieved by the average person if he or she doesn't have to work an excessive number of hours (Helvetius set the appropriate level at 7 hours a day) and has a reasonable level of basic security. Such a guarantee, Helvetius implies, can be achieved with a degree of redistribution far less than that required for complete material equality. Is it really possible to reach such an outcome with a modest amount of tinkering?

Smith, too, in the *Theory of Moral Sentiments* makes clear that material riches produce little in genuine happiness for those who own them:

> [Power and riches] are immense fabrics, which it requires the labour of a life to raise, which threaten every moment to overwhelm the person that dwells in them, and which while they stand, though they may save him from some smaller inconveniencies, can protect him from none of the severer inclemencies of the season. They keep off the summer shower, not the winter storm, but leave him always as much, and sometimes more exposed than before, to anxiety, to fear, and to sorrow; to diseases, to danger, and to death. (IV, I, 8)

For Smith, riches work as a powerful incentive to unleash human energy; they drive accumulation and benefit society as a whole. But riches are not a source of happiness. Instead, as Jerry Evensky (2005) has argued, Smith looks forward to a state of "secure tranquility" for the mass of workers, a state brought on by the extended process of capital accumulation. But Smith stops short of a direct redistribution to achieve such a state.

How appealing, then, is Helvetius's solution? For Helvetius, a modest degree of redistribution can form the basis for a considerable equalization of happiness. It would seem this is an attractive way out for Smith, Bentham, J. S. Mill, and other "analytical egalitarians." If most of us can be quite happy with a modest amount, then proposals for redistribution from the well-off can be considerably more modest than the draconian plans for achieving absolute equality of resources and/or income. Perhaps it is just such redistribution that we should take as a key component of analytical egalitarianism? Of course, for the most part the classicals, with the exception of Helvetius, don't elaborate on such a position. While we might wonder what held them back, the happiness gambit suggests a historically relevant and plausible solution to the implications of analytical egalitarianism.

G. Equality Before God

Peart and Levy make much of the alliance between classical political economy and mid-19th-century evangelicals. A key element of the alliance was the shared conviction that each of us is of equal significance in the eyes of God. From a very different tradition, that of Anglo-Catholicism, G. K. Chesterton (1933) expressed this equality in describing a convention of Medal of Honor winners. In such a congregation, who would attempt to find deep meaning in differentiating among the delegates based on their height or weight? Such differences must be largely irrelevant to their shared bravery.

In the here and now, what does such equality imply? The utilitarians had a clear answer. Starting with William Paley (1786), a religious utilitarian and contemporary of Bentham, the greatest good principle was essentially equivalent to the Golden Rule.[8] Adam Smith had also seen the Golden Rule as the practical extension of the sympathy we instinctively feel for others. And similarly, John Stuart Mill linked the greatest good principle to the Golden Rule.

Peart and Levy's construct of analytical egalitarianism echoes these identifications. As such, it opens itself to the massive range of theological debates on the social demands made by the Golden Rule. Peart and Levy make clear that evangelicals in the mid-19th century were sure that the Golden Rule required hostility to slavery throughout the world. Less clear is what the Golden Rule had to say about the treatment of the Irish or public support for education. With the exception of the issue of race, the religious dimension of analytical equality remains rhetorical here, rather than firmly rooted.

H. Political Equality

The early liberal argument for political equality was a defensive one. That argument rested on the perceived need to prevent the state and well-connected elites from encroaching on the common people. Democracy was meant to ensure that the state did not ravish its own citizens. Again, one hardly needs to accept the radical proposition of a fundamental equality of talents or intelligence to justify this limited notion of political equality. People can pretty easily tell when they are being ravished.

But Peart and Levy's discussion of political equality is out after bigger game than a simple negative check on excess and exploitation. Indeed, they seek a much stronger interpretation of the wisdom of the people. The argument borders on mystic and is supported by only a thin piece of empiricism. Peart and Levy are convinced that the aggregated political judgment of the people will often, indeed most often does, prove wiser than that of the experts. In support of this position, they discuss at some length the statistical work of Galton on fairgoers' ability to guess the weight of large bulls. While the variance of guesses was large, the central tendency tended to be right on target.[9] This same observation lies behind the current widespread interest in opinion markets for predicting elections, business cycle turns, and the like.

This is an interesting line of thought, and the authors promise to produce a good deal more work on the theme. Still, in doing so, they should be careful not to fetishize the agglomeration of public wisdom. The question is not whether this or that aggregation rule performs better in various contests. The question is how best, in a large democracy, to organize and engage in open discussion, how best to expose ideas and proposals to broad constructive criticism and active deliberation. Writing such a constitution is far trickier than picking a voting rule. In particular, it may well involve a commitment to a good deal of material redistribution, in the form of both education and substantive material support. At least that is my suspicion.

Peart and Levy's endorsement of the popular will seems to echo William Buckley's famous assertion that he would rather be governed by the first hundred names in the Boston phone book than by the faculty of Harvard University. Throughout *Vanity*, there is a strong distrust of experts.[10] Now, clearly, the authors can't be hostile to all experts. We go to a doctor for medical advice, to a lawyer for legal advice, to an accountant for help with our taxes. These are experts, and I'm pretty sure that Peart and Levy would not trust their medical care to a poll of the first hundred names in the Boston phone book.

The problem that worries them is not that as a result of the division of labor some of us know about some things and others about other things. Rather, they fear "experts" empowered by a state (does it matter if it's a democratic state?) to work their will on the rest of us. Now even here, presumably, the authors accept some role for expertise. The public health department exerts strong powers of quarantine, presumably with Peart and Levy's approval. As a result of these powers, hopefully many will live who otherwise would have died, although we recognize that a few will die who otherwise might have lived. We trust to the experts to make these decisions. Under some circumstances we want experts, even government experts, to make life-and-death choices. Experts can and do have strong public support when their skills are usefully employed. If the economic central planners could do as good a job as the public health department, not a few of us would find central planning quite reasonable.

But there are any number of experts whom we don't particularly trust. Peart and Levy make much of the role of experts in the eugenics movement that commanded so much attention from economists in the late 19th and early 20th centuries. With 20-20 hindsight, we can identify the various eugenics "experts" as vain and racist. They are an easy target. But I don't think the issue raised by eugenics is one of "experts" versus the people. A democratic canvass using medians or any other measure of the German public in the late 1930s probably would have found a considerable majority in favor of the eugenics program of the Third Reich. Indeed, most Germans of the time probably saw that eugenics program as a legitimate extension of public health measures. If the experts were misguided, the majority of the people were racists. Not much to choose here. Both "expert" knowledge and folk knowledge can be wrong.

Acknowledging the all-too-common existence of such frightful error is a beginning. It is not difficult to imagine situations in which much gain for many people may be achieved at considerable loss to relatively few. Some argue that such problems are impossible to solve within a utilitarian framework (see Wiggins 2006). I disagree. Utilitarianism seems the only reasonable way to approach such situations. The point, though, is to avoid both the "vanity of the philosopher" and the "prejudice" of the people in favor of a fuller effort both by experts and the populous to understand the consequences of our actions.

There are many things wrong with the broad modernist agenda in the social and behavioral sciences. And, certainly, modernism coupled with utilitarianism has contributed to the arrogance of not a few philosophers. But one key element of the modern program must continue to command respect. A central message of modernism holds that we must dig under the surface to find the truth. Despite their own impressive efforts at unearthing the truth, Peart and Levy seem on the edge of denying the need for such hard digging. At points in *Vanity* it seems that if we sample the populace and appropriately add up the results we can answer a surprising range of questions.

But aggregation is hardly the same as digging. While glorifying the scientific content of economics and related social disciplines is a serious and too common mistake, so is trivializing that content. When experts work as hired guns, they must be questioned closely. When they pick sides and argue incessantly, they may do little to advance the usefulness of knowledge. Expertise needs to be reproducible. Economists and others need to work more diligently at applying serious scientific criteria to solving problems and to acknowledge more explicitly what remains unknown.[11] At the same time, political equality must build on a general political maturity that can both identify charlatanism (even when ostensibly supportive of otherwise attractive political positions) and tolerate ambiguity.

Where does such political maturity come from? The utilitarians and classical economists had a clear answer: It comes from education. For Helvetius through John Stuart Mill, education yields benefits not only to individuals, but to society in the form of political maturity. Linked to a faith in at least a rough equality of intelligence, the call for education takes a key place in utilitarian prescription. But this brings us full circle. Without a measure of material redistribution, what can we hope from extensive efforts at public education? Analytical egalitarianism, again, seems to stop short in its own argument.

III

Classical Equality

The "Vanity of the Philosopher" is a stimulating and multifaceted work. It presents a huge range of material and will certainly generate considerable discussion, as the set of papers in this collection already suggests. Peart and Levy are in favor of more medians, more sympathy, and more democracy. I think I know what they mean by each of these, and readily admit to sharing all three preferences. Peart and Levy are also for analytical egalitarianism. Equality is trickier ground.

The story told in *Vanity* is one of a fall from grace. Classical political economy together with evangelical Christianity had formed a mid-19th-century alliance deeply committed to the principles of equality. In the late 19th century and early 20th century, economics became sullied by an obsession with eugenics and notions of survival of the fittest. Economics gave up analytical egalitarianism, accepted analytical hierarchicalism, and lost its classical soul.

The story is fascinating, and is richly illustrated by Peart and Levy's impressive knowledge of the intellectual history of the period. But the starting point, the analytical egalitarianism of the classical school, viewed through the embarrassment of economics' flirtation with eugenics, can take on exaggerated heroic proportions. The analytical egalitarianism of Smith and John Stuart Mill, while rich with goodwill, stopped considerably short of the radical proposals that might seem to follow from its own principles. It was the key to classical opposition to slavery and the base for the classical call for universal education. It went no further. Given its fundamental axioms, analytical equality combined with a utilitarian political philosophy would seem to call for a more radical commitment to material redistribution.

Notes

1. See, for example, Stefan Gosepath's thoughtful effort in the entry "Equality" in the *Stanford Encyclopedia of Philosophy* (2007). Gosepath quotes Ronald Dworkin in identifying equality as a "contested concept" (Dworkin 2000: 2).

2. This also seems to be Dworkin's concern in arguing the logic of his own definition (Dworkin 2000: 3).

3. An earlier generation of economists was still well aware of Smith's radical position on equality of capacity. See, for example, Tawney (1931), who attributes Smith's views to the "psychological and political theory of the age between 1750 and 1850" (1931: 47). Tawney, even in the midst of an attack on the inequality of industrial society, goes on to suggest the classical theory "greatly underestimated the significance of inherited qualities and greatly overestimated the plasticity of human nature."

4. The early utilitarian Helvetius (1772) held men to be born with an equality of intelligence: "Man is born ignorant; he is not born a fool" (1772: 6). Nurture in the form of superstition and miseducation makes men stupid. Smith, while in Paris in 1766, had personal contact with Helvetius. Helvetius also had a strong effect on Bentham, who awakened to the truths of utilitarianism while reading him.

5. Or that he did so based on utilitarian principles. In this respect, notice that Levy (1995), at least, is temped to see Smith as a utilitarian. For a recent philosophical discussion of Smith's utilitarianism and nonutilitarianism, see Fleischacker (2004). Fleischacker isn't as different from Levy as one might first imagine.

6. In many ways this is the gist of Jerry Evensky's recent book, *Adam Smith's Moral Philosophy* (2005). As will become clear below, I think Evensky lets Smith off a bit easily.

7. Peart and Levy (2005: 212–216). Peart and Levy enlist both J. S. Mill and the early Spencer as supporters of Smith's position and in opposition to the later work of Edgeworth.

8. Religious utilitarians like Paley added the pleasures and pains of the hereafter to their cost-benefit analysis.

9. Peart and Levy credit Galton with considerable moral character for supporting this position even though he was a strong believer in the wide dispersion of native ability.

10. The authors seem to identify "hierarchy" with the claim of expertise. This identification might well be questioned. Hierarchical authority in society is often exercised in opposition to or without regard to the suggestion of experts.

11. The work of Mellers, Hertwig, and Kahneman (2001) on adversarial cooperation has a nice sound of goodwill to it. Might help to rebuild public confidence. There is too much of the con in current public policy debates.

References

Bentham, Jeremy. ([1787] 1818). *Defence of Usury*, 4th ed. London: Payne and Foss. Available at Library of Economics and Liberty: http://www.econlib.org/library/Bentham/bnthUs1.html.

Chesterton, G. K. (1933). *St. Thomas Aquinas*. New York: Sheed & Ward.

Dworkin, Ronald. (2000). *Sovereign Virtue: The Theory and Practice of Equality*. Cambridge: Harvard University Press.

Evensky, Jerry. (2005). *Adam Smith's Moral Philosophy: A Historical and Contemporary Perspective on Markets, Law, Ethics, and Culture*. Cambridge: Cambridge University Press.

Fitzhugh, George. (1854). *Sociology for the South: Or, The Failure of Free Society*. Richmond, VA: A. Morris.

Fleischacker, Samuel. (2004). *On Adam Smith's* Wealth of Nations*: A Philosophical Companion*. Princeton, NJ: Princeton University Press.

Gosepath, Stefan. (2007). "Equality." *The Stanford Encyclopedia of Philosophy*, Fall 2007 Edition. Ed. Edward N. Zalta. Available at http://plato.stanford.edu/archives/fall2007/entries/equality/.

Harris, Joseph. (1757). *Essay Upon Money and Coins*. Wakefield, UK: S. R. Publishers. Rpt. 1970 New York: Johnson Reprint Corp.

Helvetius, Claude Adrien. (1772). *A Treatise on Man; His Intellectual Faculties and His Education*. Trans. W. Hooper. New York: Burt Franklin [1969 reprint of 1810 edition].

Hume, David. ([1752] 1987). *Essays, Moral, Political, and Literary*. Ed. Eugene F. Miller. Indianapolis, IN: Liberty Fund, Inc. Available at http://www.econlib.org/library/LFBooks/Hume/hmMPL35.html.

Levy, David. (2001). *How the Dismal Science Got Its Name: Classical Economics and the Ur-Text of Racial Politics*. Ann Arbor: University of Michigan Press.

——. (1995). "The Partial Spectator in the *Wealth of Nations*: A Robust Utilitarianism." *European Journal of the History of Economic Thought* 2: 299–326.

Mellers, Barbara, Ralph Hertwig, and Daniel Kahneman. (2001). "Do Frequency Representations Eliminate Conjectural Effects? An Exercise in Adversarial Collaboration." *Psychological Science* 12(4): 269–275.

Miller, William. (1966). "The Economics of Education in English Classical Economics." *Southern Economics Journal* 32(3): 294–309.

Paley, William. ([1786] 2002). *The Principles of Moral and Political Philosophy.* Indianapolis, IN: Liberty Fund.

Peart, Sandra, and David Levy. (2005). *The "Vanity of the Philosopher": From Equality to Hierarchy in Post-Classical Economics.* Ann Arbor: University of Michigan Press.

Persky, Joseph. (2006). "On the Conservatism of Radical Sympathy: Security, Equality and the Greatest Good of the Greatest Number," mimeo.

Smith, Adam. ([1776] 1937). *An Inquiry into the Nature and Causes of the Wealth of Nations.* Ed. Edwin Cannan. New York: Modern Library.

Tawney, R. H. (1931). *Equality.* London: George Allen & Unwin.

Wiggins, David. (2006). *Ethics: Twelve Lectures on the Philosophy of Morality.* Cambridge: Harvard University Press.

Thinking About Analytical Egalitarianism

By David M. Levy and Sandra J. Peart*

On the occasion of this symposium we have taken the opportunity to revisit what we have called analytical egalitarianism with the comments of our colleagues in mind. Before we began *Vanity*, we were intrigued by two discussions about types of agents. The early work in public choice by James Buchanan and Gordon Tullock (1962) held emphatically that agents in market transactions were no different from agents in political transactions. The socialist calculation debate illustrates the danger of supposing that planners could be counted on to act in the public interest (Levy and Peart 2008d). Folk wisdom in economics also suggests that economists are not exclusively truth-seeking agents. "Torture the data until it confesses" is one expression of this attitude attributed to Ronald Coase. Edward Leamer's "Let's Take the 'Con' Out of Econometrics" (1983) is perhaps the best known statement of this.

What we've tried to do in *Vanity* is to recover a tradition. We offer two requirements for analytical egalitarianism (AE). First, differences among types of agents in an AE model are endogenous to the model. Second, the theorist who creates an AE model is the same type as the agents in the model. We prefer this characterization to the older "nature" and "nurture" terminology that Kevin Hoover suggests because our characterization explicitly takes the modeler into consideration.

These conditions are in keeping with our reading of Adam Smith's practice. Philosophers and porters are very different types of people, but all differences are explained by the working out of the model that Smith describes. As a result of the division of labor, philosophers offer their insights in exchange for other goods. The question is how to

*David Levy is at the Center for Study of Public Choice, George Mason University, Fairfax, VA 22030. Sandra Peart is Dean, Jepson School of Leadership Studies, University of Richmond, Richmond, VA 23173.

American Journal of Economics and Sociology, Vol. 67, No. 3 (July, 2008).

keep the benefits of the division of labor (expertise) without giving experts the power to exploit the deference accorded to them.

AE is thus a characterization of types of agents in a model and the relationship between the model and the modeler. For many problems we can adopt the familiar "as-if" move. Indeed, as we are unlikely to work with models in which height is an interesting variable, we can afford to laugh at ourselves and the AE supposition. The hard question is, what do we do when both AE and laughter fail? We start where we began, with Smith. In *Theory of Moral Semtiments,* Smith writes about those whose natural endowment is a cause for regret ("idiots"), and he distinguishes between those who accept their inferiority and those who out of pride struggle to see themselves as the equal of others ([1759] 1982 VI iii: 49).

An example closer to the material in *Vanity* was provided by A. C. Pigou in the chapter from *Economics of Welfare* entitled "The Quality of People." The chapter starts by asking the question of whether the claims of the biostatisticians concerning genetic differences among people, which Pigou accepts as true, imply that the traditional imperative of economic utilitarians toward income equality would be reversed. What if some people were more capable of happiness than others? Pigou developed the intriguing line of argument that if the capacity for happiness were endogenous, equalization of income in itself would help equalize the capacity for happiness ([1932] 1950: 106–122). This aspect of Pigou's work is apparent also in John Rawls's commentary on the utilitarian tradition (Rawls [1971] 1999: 222; Levy and Peart 2007). The endogeneity of capacity may help explain why the traditional measure of cognitive capacity—performance on IQ tests—has been growing rather markedly over time (Flynn 2007).

We now turn to some specific points for discussion. For ease of exposition, we'll divide the responses into two sections; first, those that are correct and, second, those that warrant further discussion.

I

Criticisms That Are Precisely Right

DAN HAMMOND RIGHTLY HIGHLIGHTS what is in the "shadows" of *Vanity*: religion. While the Irish as a people are targets of those who favored hierarchy, behind the people often lurked their religious beliefs. And much of the discussion centered on the Catholic Church in Ireland. Moreover, we agree that it is important to make our commitments clear to the reader.

Andrew Farrant is correct to emphasize James Mill's *Analysis of the Phenomena of the Human Mind* (James Mill [1829] 1869; J. S. Mill 1869). We've begun to remedy that defect. Not only do we now have a much clearer picture of sympathy in the 19th-century discussions—from Smith to Dugald Stewart to James and John Stuart Mill—but we've also acquired a new understanding of Smith's emphasis on motivation by praiseworthiness. Motivation by praiseworthiness may answer the question of how the two great reforms Smith advocated, free trade and abolition of slavery, could be effected despite Smith's pessimism to the contrary (Peart and Levy 2007). If a world without special privilege becomes a norm and action to move the world closer to the norm is a praiseworthy act, then the desire to behave in a praiseworthy manner can change the world.

Kevin Hoover and Farrant are correct that we did not fully appreciate the depth of Lionel Robbins's AE. Here again our work with Rawls has been helpful (Levy and Peart 2007; Peart and Levy 2008a). Rawls reads Robbins as offering the best form of utilitarianism (Rawls [1971] 1999: 222). Robbins's *Nature and Significance,* as well as his essays on Philip Wicksteed that bracket the publication of *Nature and Significance,* clarify that Robbins's strictures on the judgment of capacity for happiness are not meant as a prohibition to make such judgments but instead to clarify that such judgments obtain warrant because they are what people do. Economists use the judgment of equal capacity because the world they model is founded on such a convention.

Charles McCann is correct that AE does not require a commitment to fixed types; what is required is that the type be endogenous. Herbert Spencer is the fascinating example of a thinker for whom "nature" is endogenous. The secondary literature still "credits" Spencer with eugenics, a bizarre reading. He was clearly a major influence on A. R. Wallace, whose doctrine that sympathy interferes with natural selection was the target of the first eugenic thinking. Agents evolve at the speed of language acquisition for Spencer, but the evolution is endogenous. What happens, he asks, if Irish people move to America but because of discrimination they do not intermarry with the earlier immigrants? In three generations they become fully American. For Spencer, the difference among agents at any moment in time is small enough to neglect (Levy and Peart 2008e). There is still a great deal of historical work to be done to be able to dehomogenize Spencer, Wallace, and Darwin. Publishing the full text of Darwin's letter on contraception (Peart and Levy 2008b) might be a step in this direction.

II

Topics for Further Discussion

McCann worries whether we have been too easy on Mill and his view of marriage. Here, perhaps what needs to be worked out is Mill's view of the changing incentives to restrain marriage under capitalism and socialism. Under socialism, as Malthus foretold, disapprobation and social pressure were the only means to restrain reckless marriage (Levy and Peart 2008d; Peart and Levy 2008b). What McCann sees as paternalism in Mill might instead be Mill's argument that a society with material incentives can avoid this fate.

Joseph Persky asks about AE, moral equality, and political equality. Is it consistent to adhere to AE without going to E? The easy answer is to say that AE is about models while E is about the world, and let it go at that. Since we have argued that models have motivational force, however, such an answer is unsatisfactory. As a possible approach to the problem, we propose a mapping from what we suppose about our agents to our commitments. Thus, if our agents are supposedly sympathetic, this sympathy characterizes our commitments and motivates action to support group activity.

Supposing that the sympathetic gradient is a function of social distance, the question arises of what the social distance is among the agents in our models and between the modelers and the agents. If the modeler is outside the model and there are no exogenous differences among types in the model, then it is reasonable to suppose that the modeler is the same social distance from the agents in the model. This would argue for an impartial egalitarianism. Perhaps this is why utilitarian economists are attracted to egalitarianism.

In the world beyond the model, agents who are modeled have lives of their own. The agents are not at equal social distance from each other, so there is no reason to believe that they themselves will be egalitarians. Like Rawls, they may very well be more attracted to Knight's fair games with a generous safety net than to the substantial equality of Pigou's utilitarianism. We fall back, once again, to Smith, who defended a modification of stoicism in which differences in social distance were real and had motivational force (Levy and Peart 2008a, 2008c). Nonetheless, to combat the temptation of partiality, our agents may need to learn to see themselves from outside. Smith explains that this is the great power of stoicism (Peart and Levy 2007; Levy and Peart 2008a).

Kevin Hoover queries whether we have succumbed to the vanity against which we warn, attempting to win an argument by telling those who disagree to shut up. In particular, how do studies that underlie eugenics differ from the analysis of the impact of abortion on crime or politics?

In our view, eugenics is the oldest form of central planning. As such, the eugenic expert faces a temptation. The expert must distinguish inferior from superior people. The policy proposals that attempt to increase the proportion of superior people in the population depend upon this ability to make the distinction between superior and inferior. If the distinction can be made, the rewards to the expert will be greater than if the distinction cannot be made. This asymmetry in rewards may tempt the expert to find distinctions when none exists. We see no such asymmetry in rewards to research on the impact of abortion on crime or politics.

The temptation of the expert is a reason why we prefer to get away from the old "nature" versus "nurture" terminology that seems to suggest that the expert is not part of the analysis. We want to put the expert in the model and ask whether truth—or its statistical equivalent—is incentive-compatible. The public at large is starting to learn a great deal about the temptation of experts from recent events on Wall Street. The incentives of rating agencies to produce one result rather than another are now becoming clear. Once the temptation is understood, then the incentives may be changed to help make expertise more transparent (Peart and Levy 2008c; Levy and Peart 2008b;

Zelder 2008; Houser 2008). If nothing else, AE might serve as a benchmark: Do we trust the conclusions of a model once we have taken the interests of the modeler into account?

References

Buchanan, James M., and Gordon Tullock. (1962). *Calculus of Consent.* Ann Arbor: University of Michigan.

Flynn, James R. (2007). *What is Intelligence? Beyond the Flynn Effect.* Cambridge: Cambridge University Press.

Houser, Daniel. (2008). "Experimental Ethics." *Eastern Economic Journal* 34(1): 126–128.

Leamer, Edward. (1983). "Lets Take the Con Out of Econometrics." *American Economic Review* 73(March): 31–43.

Levy, David M., and Sandra J. Peart. (2007). "Efficiency or a 'Fair' Game: John Rawls *Contra* Lionel Robbins." Presented at the conference "The Nature and Significance of Economic Science at 75," London.

——. (2008a). "The Evils of Independence: Stoics Sources for Adam Smith." *Adam Smith Review* forthcoming.

——. (2008b). "Inducing Greater Transparency: Towards the Establishment of Ethical Rules for Econometrics" *Eastern Economic Journal* 34(1): 103–114.

——. (2008c). "The Place of Factions." In *Elgar Companion to Adam Smith.* Ed. Jeffrey Young. Aldershot, UK: Edward: Elgar.

——. (2008d). "The Socialist Calculation Debate." In *New Palgrave Dictionary of Economics*, 2nd ed. Ed. Steven Durlauf and Lawrence Blume. London: Palgrave.

——. (2008e). "Sympathy, Evolution and *The Economist.*" *Advances in Austrian Economics* forthcoming.

Mill, James. ([1829] 1869). *Analysis of the Phenomena of the Human Mind.* Ed. John Stuart Mill. London: Longmans, Green, Reader and Dyer.

Mill, John Stuart. (1869). "Notes." In *Analysis of the Phenomena of the Human Mind.* Ed. John Stuart Mill. London: Longmans, Green, Reader and Dyer.

Peart, Sandra J., and David M. Levy. (2007). "Adam Smith on Leadership, Equity and the 'Distinction of Ranks'." Presented at the International Leadership Association, Vancouver.

——. (2008a). "The Buchanan–Rawls Correspondence." In *Street Porter and the Philosopher: Conversations on Analytical Egalitarianism.* Ed. Sandra J. Peart and David M. Levy. Ann Arbor: University of Michigan Press.

——. (2008b). "Darwin's Unpublished Letter at the Bradlaugh-Besant Trial: A Question of Divided Expert Judgment." *European Journal of Political Economy* forthcoming.

——. (2008c). "Introduction to the Symposium." *Eastern Economic Journal* 34(1): 101–102.

Pigou, A. C. ([1932] 1950). *The Economics of Welfare*, 4th ed. London: Macmillan.

Rawls, John. ([1971] 1999). "Justice as Reciprocity." In *Collected Papers*. Ed. Samuel Freeman. Cambridge: Harvard University Press.

Smith, Adam. ([1759] 1982). *The Theory of Moral Sentiments*. Eds. D. D. Raphael and A. L. Macfie. Indianapolis, IN: Liberty Fund.

Zelder, Martin. (2008.) "Why Hasn't the Con Been Taken Out of Econometrics?" *Eastern Economic Journal* 34(1): 115–125.

III. ANTHROPOSOCIOLOGY

Social Anthropology in Economic Literature at the End of the 19th Century

Eugenic and Racial Explanations of Inequality

By TERENZIO MACCABELLI*

ABSTRACT. At the end of the 19th century, Georges Vacher de Lapouge and Otto Ammon founded a school of thought denominated "social anthropology" or "anthropo-sociology," aimed at placing racism on a scientific basis. Their intent was to create a new discipline into which the themes of biological heredity, natural selection, social stratification, and political organization were to converge. This paper intends to demonstrate the wide resonance that anthroposociology had in the economic literature, analyzing the thought of authors such as Carlos C. Closson, Vilfredo Pareto, and Thorstein Veblen. A particular focus will be on the racial and eugenic arguments used as explanation of social and economic inequality.

Social anthropology is today considered a flourishing branch of human knowledge, promoted all over the world by specialist journals and scientific institutions.[1] However, the social anthropology dealt with in these pages has little to do with this reality. The object of this research is the thought of authors such as the Frenchman Georges Vacher de Lapouge and the German Otto Ammon, who, in the second

*Terenzio Maccabelli is an Assistant Professor at the University of Brescia, where he teaches History of Economic Thought; e-mail: maccabel@eco.unibs.it At present his research activity mainly deals with the problem of inequality and, in particular, the "Pareto law." He would like to thank the anonymous readers of *AJES* and Laurence Moss for their helpful comments. He is extremely grateful to Francesco Asso, Fabrizio Bientinesi, Luca Fiorito, and Tiziana Foresti for their bibliographical support. None of them are, of course, responsible for any errors.

American Journal of Economics and Sociology, Vol. 67, No. 3 (July, 2008).

half of the 19th century promoted a new discipline, with an explicitly racist and eugenic content that they initially called "social anthropology" and later named "anthro[po]-sociology." In the last decades of the 19th century, this discipline had vast resonance in European and American culture. Its fortunes were obscured with the beginning of the new century, but, however, were strongly rooted in Germany, where it became an incubator for the National Socialist eugenic projects (Mosse 1978). This work intends to demonstrate the wide resonance that this interpretation of social anthropology also had in the economic literature of the period.

One of the inducements to undertaking this research is the increasing interest shown by historians of economic thought in the relationship between eugenics, racism, and political economy. Numerous contributions on the question recently have been published,[2] nonetheless, neglecting the thought of the so-called school of social anthropologists. Much more studied instead is the reception given to the theories of Lapouge and Ammon in the sociological field,[3] where the parabola of the doctrine has been highlighted; after sudden popularity, sanctioned by the space dedicated to it in the major sociological reviews, social anthropology was progressively ousted from the scientific sociological panorama following its scientific delegitimization. In this paper we propose an analogous reconstruction from the economic standpoint. In analyzing the spread of anthroposociological theories in the economic field, major emphasis will be laid on the eugenic and racial explanations for social stratification, which constitute the heart of such a doctrine.

The paper is organized as follows. The first section is dedicated to a brief presentation of the social theories of Georges Vacher de Lapouge and Otto Ammon, the founders of the school of social anthropology. The second section comprises an overview of the articles that the most authoritative economic reviews of the period, principally the *Quarterly Journal of Economics* and the *Journal of Political Economy*, dedicated to the school, chiefly through the agency of Carlos C. Closson, an indefatigable popularizer of the theories of Lapouge and Ammon in the English-speaking world. The third section of the paper enters into the principal question; that is, the explanation offered by the social anthropologists for economic and social inequality. The fourth section then discusses the influence of social anthropology on Vilfredo Pareto, one of the most authoritative scholars of social hierarchies, and shows how responsive he was to the topics discussed by Lapouge and Ammon. In the fifth part, the paper will illustrate the growth of criticisms of social anthropology and how it was ostracized with the coming of the new century, both by economists and by sociologists. In the last part will be discussed the role of Thorstein Veblen in the spreading of social anthropology in economic literature, before of its epilogue.

I

The School of Social Anthropology: Georges Vacher de Lapouge and Otto Ammon

AMMON AND Lapouge are authors who are seldom quoted in the history of social science.[4] At the turn of the century, however, their fame was notable and their writings were diffused by the journals of various scientific disciplines. The two scholars were, in truth, difficult to classify from a strictly disciplinary point of view, given that they operated in a grey area covering craniometry, anthropology, sociology, and economics. Their scientific aims were in fact intended to create of a new branch of knowledge, for which they coined the term "social anthropology" or "anthropo-sociology."[5]

Each declared himself to be the founder of the new discipline, stating a precise period for its inception. As Lapouge wrote (1897: 57), "the establishment of anthro-sociology as a distinct branch of investigation dates from my lectures at the University of Montpellier (1886–1892) and from the publication by Ammon of his research on the conscripts of Baden." Born in 1854, Lapouge became librarian at the University of Montpellier in 1886, after he failed to receive a university chair. His major works—*Les sélections sociales* (1896), *L'aryen, son rôle social* (1899), and *Race et milieu social* (1909)—are based on his lectures at Montpellier, where "he taught a variety of courses, conducted population survey, compiling craniometric data and publishing a flurry of articles." During the late 1880s he wrote in the *Revue d'anthropologie,* but in 1895 the *Revue* "ceased to accepting his contributions." In 1893 he left Montpellier and became a librarian in

Rennes. After 1900, his writings were published mainly in Germany, for the most part in the *Politisch-anthropologische Revue*, the racist journal founded and financed by Ludwig Woltmann. In his last years, he became the *maître à penser* of the official race theorists of National Socialism. "He died in 1936, virtually ignored by French intellectual circles."[6]

There is scarce biographical information on Otto Ammon (1842–1916). An engineer and journalist, he became one of the most influential racial anthropologists of his day, greatly influencing the move toward National Socialism in Germany (Gasman 1971: 148). Like Lapouge, he never held an academic post (Drouard 2005: 15). "His first contribution came as an unexpected result of an investigation of the army recruits of Baden," which he carried out in 1890 as secretary to the anthropological commission of the archaeological society in Karlsruhe. Among many other things, "he found that there was a difference between rural and urban populations as regards hair-color, eye-color and head-form" (Hankins 1931: 110). The outcome of this research was his monographs (1890 and 1893) "that established his anthropometric credentials" (Llobera 2003: 105). His major work was *Die Gesellschaftsordnung und ihre naturlichen Grundlagen* (*The Social Order and Its Natural Bases*), published in 1895 and translated into French in 1900. Albeit Lapouge designated Ammon as co-founder of his school of anthroposociology, the German writer always regarded Lapouge "as a student regards his master."[7] Together with Lapouge, Ammon was one of the most fervent supporters of Social Darwinism, spreading the idea that society, like nature itself, is subject to universal laws of natural selection and that the social sciences could not disregard Darwin's recent discoveries in the biological field.[8]

Social Darwinism appears in fact to be the cultural context in which to set Ammon and Lapouge's anthroposociology, and it is therefore hardly surprising that almost all of those who interpret their work characterize them as "Social Darwinists."[9] This appellation, while undoubtedly true, nonetheless contains some snares. "Social Darwinism" is indeed one of the most abused terms in the history of social sciences and also one of the most ambiguous. As Hawkins (1997: 32) has pointed out, " 'Darwinism' was not a fixed entity" and "there are a series of indeterminacies surrounding some of its elements." The question whether Darwin himself was a Social Darwinist is not a trivial one. Problems arise, for instance, in looking at the historical relation between Social Darwinism and eugenics, as "it was possible to support one and not the other" (Hawkins 1997: 6). This was not the case for the anthroposociological theories, which postulated a very close association between Social Darwinism and eugenics. Furthermore, not all Social Darwinists followed the racist path, while Lapouge's and Ammon's Social Darwinism was indeed a theory with an explicit racist content. This ambiguity makes it difficult to evaluate the real intellectual debt of Ammon and Lapouge to Darwin.

In order to understand the cultural matrices of social anthropology, it is necessary to begin with the radical changes that took place in the social sciences towards the middle of the 19th century, when the doctrine of social homogeneity was abandoned. Authors as diverse as Burke, Rousseau, or Smith, to cite but a few examples, held that the natural equality of human beings was an irrefutable principle. Inequality—both by those who condemned it and by those who legitimized it—was in fact considered by all as a socioeconomic fact, the product of historical evolution and of the social institutions.[10] Around the middle of the 19th century, instead, as the anthropology historian Marvin Harris has observed, there was no more "self evident truth" than the fact that all men were created *unequal.*[11] Inequality among individuals began to be held to be an objective and natural fact of a biological order as, indeed, was "the 'fact' that certain races were superior to others" (Hawkins 1997: 184). Toward the middle of the century, these ideas were widespread among the European intelligentsia, thanks above all to the literary fortune of the ideas vulgarized by Gobineau on the inequality and on the hierarchy of the human races (Battini 1995: 196). The theory of races formulated by Gobineau between 1853 and 1855 in his *Essay sur l'inegalité des races humaines* is without doubt an important juncture, as is recognized by the founders, for the birth of social anthropology.[12]

Ammon and Lapouge nonetheless make a leap in quality, in respect to Gobineau, in trying to give racism a scientific basis.[13] The distinction between races is in fact anchored on quantifiable empirical data, namely, the cephalic index.[14] Both Ammon and Lapouge used it to divide the European population into three fundamental racial types:

the *Homo Europaeus*, the *Homo Alpinus,* and the *Homo Mediterraneus. Homo Europaeus* had a lower cephalic index—that is dolichocephalic, with a long narrow skull—and is normally tall, light-skinned, and blond-haired; *Homo Alpinus* is instead brachycephalic, dark, and shorter in stature; *Homo Medditeraneus*, finally, although dolichocephalic, is also darker and shorter in stature and is found exclusively in the Mediterranean basin. The principal innovation introduced by the anthroposociologists "lay in the qualitative characteristics [they] associated with these head shapes" (Hecht 1999: 4). Dolichocephalic and brachycephalic individuals would in fact have different aptitudes, predispositions, and intellects that underlay their different social and cultural performance. *Homo Europaeus*, called *aryan* by Lapouge and *teutonic* by Ammon, was supposed to be active, enterprising, and ambitious, with a marked tendency to migrate and a singular attraction toward urban life; *Homo Alpinus* was instead more static, mostly concentrated in the agricultural centers, and little inclined to change and innovation.[15] The correlation identified between the cephalic index and human capabilities, the outcome of a titanic effort in measuring and cataloguing by Ammon and Lapouge,[16] was the proof, in their opinion, of the "scientific" nature of their racism. The hierarchy of races guessed at by Gobineau would therefore have found confirmation in anthrosociological research intended to demonstrate that even the qualitative differences between individuals could be traced back to quantitative, measurable facts.[17] These ideas were then formulated into a theory of social stratification, to be dealt with later, based on the correlation between socioeconomic status and racial stock.

Besides being based on the idea "that human racial differences are real, significant and scientifically measurable" (Leonard 2003: 689), Ammon and Lapouge's anthropology also postulates that these differences can be inherited. In the wake of Galton, Ammon and Lapouge in fact maintain that "genius" and "intelligence," no less than physical features, have a biological foundation and are therefore transmittable through heredity. The concept frame for this biological determinism is that set out by Galton with the distinction between "nature" and "nurture."[18] The social context has no power to modify the individual and racial differences that derive from "innate characteristics" and from biology. The condemnation Ammon and Lapouge make of those currents of Social Darwinism that accept the Lamarckian theory of the capacity to inherit acquired characteristics springs from this. The two anthropologists believed that education and social context do not have a role in the evolution of the races, not being able to produce real effects that could be transmitted to descendants.

A further component of social anthropology is natural selection.[19] The elimination of the least fit is a fundamental process in nature, which, nonetheless, works in a flawed and more complex fashion in human societies. "Natural selection" is transformed in fact into "social selection," "in the measure in which social context exercises its influence on the natural environment."[20] Contrasts between natural selection and social selection can occur, above all, when social institutions favor the proliferation of the mediocre and hinder the reproduction of the superior individual.[21] In modern society, in particular, norms and social practices that are in contrast to the bettering of the race prevail.[22] The brachycephalic individuals, despite being intellectually and socially "inferior," have major reproductive powers and greater adaptability to social norms. From this derives the possibility that the brachycephalics will end by replacing the dolichocephalics through selection of a complex order, both biological and social.[23] Lapouge invoked in this context "an anthropological analogue to Gresham's law in economics, according to which good coin was driven out of circulation by bad coin: when two races were mixed, the inferior would eventually predominate over the superior" (Hawkins 1997: 192).

The only way to invert the tendency toward biological decadence was that of eugenics.[24] Both Ammon and Lapouge came to maintain the necessity of the physical elimination of the "inferior" subjects, a task that, if not carried out by natural selection, would have to be performed by the state. The two anthropologists believed in this case that by catering to the predisposition of the inferior individual, such an aim could be easily pursued. The establishment of places particularly attractive to him or her—where alcohol could be distributed free, where vice could be spread, and libertine behavior favored—would in fact have led to a concentration of degenerate individuals who could easily be eliminated.[25] Naturally, such a project could never be carried out under a liberal organization of society, which therefore had to give place to a centralized and state-controlled model of socialism (or National Socialism), aimed at bringing into being the real principles of the social hierarchy. "Substituting current humanity with a unique and perfect race" needs, Lapouge wrote, "almost of necessity, a socialist regime"; this requires, however, overcoming the traditional concept of socialism, which has "shown itself up to now to be prevalently leveling and detrimental."[26]

These are the principal traits of Ammon and Lapouge's social anthropology, a new racist theory that European and American culture at the end of the 19th century received as "scientific," "erudite," and "revolutionary."[27] The early diffusion is, in truth, circumscribed to France and Germany, where the two anthropologists worked. The first articles by Lapouge, appearing "in the major French anthropological publications from the mid 1880s to the mid 1890s" (Schneider 1990: 62), had scarce resonance outside France, as did the German-language publications by Ammon in the early 1890s. Things changed after 1895 for two reasons: first because anthroposociological theories began to cross the confines of the discipline of anthropology, finding acceptance even in the economic and sociological journals;[28] and second because the echo of the anthroposociological doctrines was amplified thanks to their diffusion through journals in the English language. Lapouge's and Ammon's theories had by then gained "an aura of scientific respectability" (Weiss 1987: 93–94) recognized at the international level. As pointed out by Poliakov, when leafing through the journals and the publications of the time, one is convinced that anthroposociology "was very much taken seriously. Certainly Lapouge led the way, above all in Germany, where Kaiser Wilhelm II championed him as 'the only great Frenchman'; but he had advocates in other European countries too" (1971: 306).

II

Carlos C. Closson and the Spread of Social Anthropology in Economic Journals

A SIGN OF THE incredible good fortune of social anthropology is the resonance this doctrine enjoyed in the major economic reviews,[29] fed by an incessant publicity campaign conducted by Carlos C. Closson. We know very little of the meteoric academic career of Closson. He received his A.B. from Harvard University in 1892 and his A.M. in 1893.[30] After a fellowship in social science at Harvard, he was appointed instructor in Advanced Economics at the University of Chicago in 1894. He held the position until 1896[31] and, when he left, "Veblen was assigned the 'theory' courses, which were required for advanced degrees" (Dorfman [1934] 1947: 132). Closson declared himself a disciple of Lapouge, and he taught courses on "social selection" at Chicago (Hecht 2003: 122). Between 1895 and 1900, Closson's publishing activity was intense. He translated some of Ammon's and Lapouge's fundamental contributions for the English-language journals and published many essays himself on anthroposociology and social selection. But after 1900, the name of Closson suddenly vanished from any social and economic debates. His disappearance was as precipitous as the decline of what he ultimately called anthroposociology. What became of him in the new century is puzzling. We know only that he moved to Los Angeles and undertook a career as a businessman.[32] Thus it would seem that academic teaching and his scholarly activity were only a parenthesis in Closson's life.

The fortune of social anthropology in the economic journals appears in truth to be circumscribed to a limited time. The economic journals, in fact, acted as a sounding board for Ammon's and Lapouge's theories for only five years, from 1896 to 1900. With the beginning of the new century, the debate over social anthropology ceased suddenly, perhaps also due to some economists distancing themselves from it and its subsequent delegitimization.

Social anthropology held the limelight in the 1896 economic debate, with the publication of no fewer than four articles by Closson and the translation, by the same, of Ammon's work. Initially, Closson defined the new scientific approach as "the selectionist school of social sciences" (1896a: 156) in virtue of the key role played by the concept of "selection"; a few months later he began to talk of "social-anthropology" to underline the strict links between the new discipline and sociology, economics, and politics;[33] from 1897, the term "anthro[po]-sociology" begins to appear, which is the term destined to consecrate the new school of Lapouge, Ammon, and Closson himself. Even among economists, social anthropology is presented as a "new" science, capable of unifying data and knowledge coming from disciplinary areas hitherto far distant. Its claim to be scientific rests on the ample use of empirical and statistical data and on its ability to enunciate universal laws deducted from that data. Therefore, the American economist underlines "the importance of the new science of anthroposociology for the whole range of the social sciences" but, above all, its repercussions in the "sociological field, from which, precisely, there derives its denomination as 'anthropo-sociology' " (1898: 254–255).

Closson's contribution, which forms an effective summary of the thoughts of Ammon and Lapouge, addresses two principal questions: one begins with the process of separation that pushes individuals possessing different characteristics in the racial sense to form distinct social groups; the other deals with social selection and its relationship with natural selection. From the former point of view, the American economist highlights how the diverse aptitudes of the human race are transformed into a division within society, both on a hierarchic plane ("disassociation by stratification") and on a geographic one ("disassociation by displacement"). In both cases, social mechanisms act and lead to the separation of dolichocephalic subjects from brachycephalic ones: in the first case, through a concentration of the former into the higher classes of society; in the second, through the migratory phenomenon, leading to a concentration of the dolichocephalic in the cities. The underlying origin of this diversity of performance is held to be a biological fact: on the one hand, the fact that *Homo Europaeus*, or dolichocephalic, "seems to be superior in general psychic ability and character" and is therefore destined "to the higher position in the process of social stratification";[34] on the other hand, the greater enterprise and aggressivity of *Homo Europaeus* would explain his major mobility compared to the brachycephalics, more given to the rural life.[35]

The second theme dealt with by Closson touches the center of one of the key concepts of social anthropology, that is, selection. The 1896 publication of Lapouge's volume entitled *Les sélections sociales* led Closson to write a well-constructed summary in the *Journal of*

Political Economy. Lapouge's disciple started by indicating the principal aims of natural selection, which consist of the elimination of degenerate individuals and the perfection of the superior elements. As has already been said, however, the complexity of human society in general generates a contrast between natural selection and social selection. In society, in fact, selection at times inverts these aims.

Closson indicates two principal forces that operate in society: the conservative force of biological heredity, which tends to perpetuate the same physical characteristics from generation to generation; and the evolutionary force of selection, which introduces changes and establishes which species have to progress and which to disappear (1896d: 453). These two forces act on the races, and the second in particular "is the great force in altering the quality of the population." This bettering, however, does not come about thanks to the action of the institutional environment but happens exclusively on the racial plane. Closson emphasizes one of the pillars of anthroposociology, that is, the sterility of "nurture": "[E]ducation has only a limited effect upon the single individual and even this effect is not in any appreciable degree transmitted to his descendants; it cannot be relied upon as the primary means of human improvement." The work of selection must therefore be such as to favor the development of the "ethnically superior elements," something that does not always happen because of the institutions and the social norms that have ended up by favoring the proliferation of the "inferior classes" (1896d: 459–462).

On this occasion Closson picks up on what truly obsesses Lapouge, that is, the danger of a deterioration of the race. Faithfully reporting the passages from Lapouge, the American economist indicates the two directions in which selection must proceed: "(1) to eliminate the degenerate, vicious and incapable elements, (2) to increase and perfect the superior element" (1896d: 565). There are naturally difficulties of a moral and social order that hinder the carrying out of a similar objective, above all, the need to impose a "social" control over the mechanisms of reproduction. Closson, however, held that such reforms were necessary "in the highest interests of human welfare." Singularly, the American economist is silent regarding the radical solutions prospected by Ammon and Lapouge, who, as we have seen, outlined the features of a National Socialist society in which the question of the suppression of "inferior" individuals became a matter of state. His conclusion is nonetheless perfectly in line with the two founders of social anthropology. Without indicating the means that would allow them to reach their objectives, he enunciates the eugenic aims set out by the social anthropologists, toward which the action of social selection should be directed:

> (1) to constitute a natural aristocracy among a given people; (2) to constitute specialized and distinct castes suited for the different branches of social works, (3) to transform a whole people in a given direction, (4) to form a universal dominant race, (5) to improve all humanity by utilizing the most perfect local types, (6) to substitute for existing humanity a single more perfect race, etc. (Closson 1896d: 465)

In the closing years of the century, the publishing activity of Closson continued to be intense (see Closson 1898, 1899a, 1899b, 1900a, 1900b, 1900c). Thanks to his numerous contributions, the American economist earned his stripes as the American representative of the school of the so-called anthroposociologists; in the economic literature of the period, that school was usually identified with the names of Ammon, Closson, and Lapouge (see Ripley 1900). We owe the English translations of the most important contributions of Ammon and Lapouge, which allowed the English-speaking public to have direct contact with the thoughts of the two anthropologists, to Closson himself.[36] We will return to these shortly, concentrating our attention above all on the concept of the social and economic hierarchies formulated by Otto Ammon. The German anthropologist's theory is in fact of a certain importance and can in part be extrapolated from the eugenic and racist context in which it was formulated. It should not surprise us that such a theory has been revived, even in recent years, centering as it does on a problem that is still today at the hub of the social and economic debate.

III

The Statistical Basis of the Social and Economic Hierarchies

TOGETHER WITH Lapouge's book from 1896 on social selection, the volume written by Ammon in 1895, *The Social Order and its Natural Bases*, is without doubt one of the most representative of the social and political philosophy of the anthroposociologists. The indefatigable Closson translated the first part of the book in the *Journal of Political Economy* in 1896,[37] integrating the writing with notes and personal paragraphs. In this essay, Ammon expounds a statistical model to explain the social and economic hierarchies that were destined to enjoy great success in the history of the social sciences.

From this perspective, Ammon is within a tradition of research that had illustrious precursors. In fact, we owe to Adolphe Quetelet and Francis Galton the earliest use of the normal law of errors as an instrument for describing the aptitudes and mental capacities of an individual. Right from his earliest book in 1835, *Sur l'homme . . . , ou essay de physique sociale*, Quetelet had presented data relative not only to the weight or height of human beings but also concerning intellectual capacity, moral traits, or specific social predispositions (for example, toward crime). Organizing the data in the form of frequency distribution, he had shown how all these characteristics revealed the typical form of the Gaussian curve, that is, symmetrically disposed around the mean value (or rather, that which in the repeated observation of any given phenomenon, for example, an astronomical one, corresponds to the correct value).[38] Quetelet interpreted such "bell-shaped" distribution of human aptitudes just as he interpreted the curve of the errors in astronomical observation. He then had recourse to these data to deduce the idea of the "average man," that is, the entity that social physics should have dealt with (Gordon 1991: 530–531).

Galton took Quetelet's discussion even further, concentrating his attention prominently on the distribution of the "qualitative" characteristics (genius, intelligence, capacity, etc.). In his *Hereditary Genius* (1869), the English statistician had in fact assumed that talent was distributed normally and that deviations from the mean followed the Gaussian law of errors (1869: xi). Availing himself of the same formula used by Quetelet,[39] he had estimated the number of persons belonging to the diverse social classes, ordered in ascending order starting from the class of "true idiots and imbeciles" up to the class of the "geniuses." The result was a symmetric representation aimed at demonstrating "the vast abundance of mediocrity (i.e., the standard of intellectual power)" and "the rarity of commanding ability" (1869: 30–31). The fact that the law of errors had one of its most evident manifestations in the human world was, in his opinion, due on the one hand to the action of natural selection and on the other to heredity. Through natural selection, the physical and intellectual characteristics that were best suited to the species ended up by becoming mean average characteristics, according to modalities very similar to those described by Quetelet in his works of social physics. Heredity contributed then to reinforcing this dynamic, operating in such a way that the talents and the aptitudes of the individuals as well as their physical characteristics were subject to the normal distribution. With his starting point as the idea that intelligence was transmissible only through heredity, Galton therefore stated the need for a new science—eugenics—that would be able to better the "mental qualities" of individuals through control over the mechanics of reproduction.

In trying to demonstrate the normal distribution of human capacities, neither Quetelet nor Galton had extended their research to the field of income distribution. Only incidentally did they mention the fact that such a distribution mirrored the personal distribution of earnings and wealth.[40] Ammon's objective was precisely that of filling this gap. On the one hand, he insisted that the intuition by Galton and Quetelet of the normal distribution of aptitude was well-founded; on the other, he set himself to demonstrating how such a distribution perfectly corresponds to the distribution of earnings and wealth.

On the first point Ammon uses a statistic artifice, considered as able to give Galton's theory even more foundation. The English scholar had assumed a priori that the distribution of aptitude was normal; Ammon instead proposed a simulation that was intended to demonstrate the statistic process that generates such a distribution, which would find confirmation also in the theories of probability. The German anthropologist imagined that the gifts of each individual concern four distinct areas: intellectual, moral, economic, and physical. He then supposed a scale of evaluation from 1 to 6 for each of these areas, in which 1 is the minimum aptitude value and 6 is the maximum. Evidently excellence will correspond to a combination of six points in all four areas (for a total of 24), while inferiority is measured by one point in each area (a total of four points).

Thanks to the theory of probability, Ammon continued, it is possible to know the type of distribution that would prevail in this hypothetical community. It is sufficient to think of dice, of the "throwing" of four data, each of which represent the intellectual, moral, economic, and physical traits of the individual. The number of all the possible combinations (6 × 6 × 6 × 6 = 1,296) is equal to the total population. The lowest probability (1/1296) will be associated both with the combination with the highest total score (24)[41] and with the combination with the lowest score (4).[42] The two cases correspond to the extremes of the social scale, that is, to the individuals who are more gifted, on the one hand, and to the those who are totally bereft of intellectual, social, and physical aptitude, on the other. An ever-increasing frequency will be found for the intermediary combinations, with the probability and the maximum frequency associated with the average value (which in the case hypothesized corresponds to the sum of 14, which can be realized by 146 different combinations). The correspondence between the statistical significance of the experiment and its economic and social relevance was, in Ammon's opinion, quite evident:

> The number of extremely high throws, as well as of extremely low ones, is comparatively small, while the average or nearly average throws appear very frequently. Directed toward our present subject, this signified that the number of men of genius and high talent is relatively small, because in accordance with the mathematical laws, the necessary combination of traits can appear only seldom. Correspondingly, the number of the weak-minded and of the imbecile is relatively small, while the number of those of about average ability is far in excess of that of the two extremes. (Ammon 1896: 209–219)

A graphic representation of such distribution would appear similar to that obtained by Galton from the Gaussian formula of errors (see Figure 1).[43] Ammon, however, considered a partial modification of the figure to be necessary to make it conform more to what he believed to be the real form of the "social pyramid."

Two important aspects stand out: in the first place, the substitution proposed by Ammon of the image of the pyramid with a figure similar to that of an "onion," considered to be more appropriate in representing the "true" form of social stratification; in the second place, the expediency of avoiding the lowest part of the curve to underline the imperfect symmetry of the social hierarchy. The solution proposed by the German anthropologist arose from his conviction that natural

Figure 1

Ammon's Social Pyramid

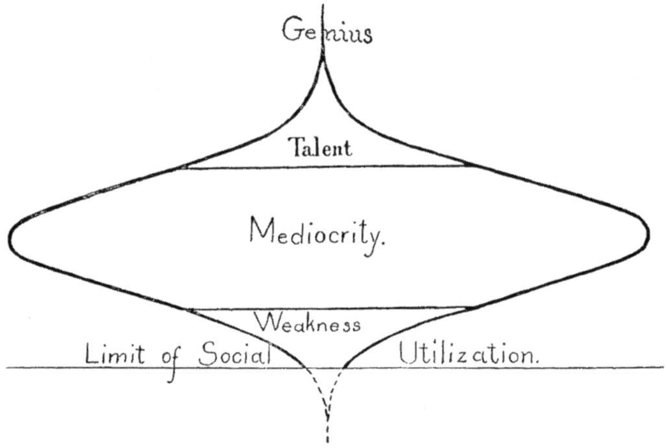

selection eliminates subjects with excessively low aptitudes, considered insufficient to survive in society.[44] The lower limit of the social hierarchy commences in fact after the line that Ammon defined as the "limit of social utilization."

Ammon's reasoning proceeded with what is perhaps the earliest attempt to demonstrate the coincidence between the "intelligence curve" and the "wealth curve," a hypothesis already outlined by Galton but without the necessary support of the statistics of incomes. Using these fiscal statistics,[45] the German anthropologist superimposed the Galtonian curve of ability on the curve of the distribution of incomes, revealing a notable correspondence between the two curves. The economic and social stratification illustrated by the statistics on incomes is therefore held to be the reflection of the diverse combinations of the aptitudes possessed by the individuals. Ammon therefore concluded that that "two truths" emerge from the comparison of the two curves:

> (1) that the form of the curve of incomes (except at the base) very nearly coincides with Galton's curve of the distribution of ability, and (2) that the income curve is not symmetrical above and below but corresponds more

nearly to what we have called the "true form of the social pyramid" . . . which like the income curve, stands upon a horizontal base line. (Ammon 1896: 226)

As we have seen, up to this point Ammon's reasoning seems to be free of racial connotations. His conceptual model decrees only the inevitability of a hierarchical organization of society without giving any indication as to the subjects who should occupy the different rungs on the social scale. We could even deduce that such distribution is governed by "chance." The fact that the number of more gifted individuals (and, as such, at the apex of the social-economic hierarchy) is limited is the exclusive outcome of the laws of statistical probability. In Ammon's concept, however, natural selection does not "play at dice," to paraphrase a celebrated aphorism. The model is therefore led back to a framework of biological determinism. The superiority of the individual at the apex of the social hierarchy is not only due to the fact that "the average ability among the higher classes is more favorable than among the lower" but also, and above all, to the fact that "this difference rests upon a hereditary base." It is not therefore a case of casual extraction by chance that assigns to each individual his or her proper stock of aptitudes, but instead the individual's descent from a bloodline with determinate biological characteristics.

The statistical model elaborated by Ammon is supported by "The Fundamental Laws of Anthropo-Sociology," as the title of the essay published by Lapouge in the *Journal of Political Economy* in 1897 tells us. The essay summarizes the principal findings by Lapouge, Ammon, and Closson, transformed into a long list of social laws. Two of these are of particular importance: (1) the law of the distribution of wealth, and (2) the law of stratification. They permit "the filling in" of the stylized representation of the social hierarchies proposed by Ammon with the different racial types.

The first law states the major quota of wealth possessed by the dolichocephalic subjects.[46] This theory is supported by Lapouge with numerous tables and correlation indexes that were meant to show the evidence of a major financial contribution in the areas where the *Homo Europeaus* or Aryan type was dominant. This is the sign, in the opinion of the French anthropologist, of the diverse "economic efficiency" of the races. The correlation indexes in particular "demonstrate the economic, and especially the commercial, superiority of *Homo Europaeus*" (Lapouge 1897: 63). Lapouge stated that the law has a universal nature when a comparison is made between dolichocephalic and brachicephalic populations. On the other hand, the existence of a correlation between the cephalic index and wealth in those countries composed almost exclusively by dolichocephalic subjects was still controversial. In such a case it would not in fact be a comparision between *Homo Europaeus* and *Homo Alpinus* but "between different degrees of dolichocephaly among the former racial element" (1897: 66). The formulation of the law with greater universality is therefore that which states "in the countries where *Homo Europaeus* predominates, wealth is in general the greatest (England, the United States, Holland, etc.)" (1897: 66).

The second law concerns social stratification. It states, independently of the degree of wealth, that dolichocephalic subjects have higher social status.[47] That is, there would seem to be a "correlation between race and social position," proven by the correspondence between the cephalic index "and the profession of the individuals concerned" (1897: 87). It is evident that if social status runs side by side with income and wealth, then the law of the distribution of wealth and the law of stratification are no less than two faces of the same coin. The superimposition of the curve of ability and the curve of incomes proposed by Ammon intends, in fact, to demonstrate, as we have seen, exactly the convergence of the economic hierarchy with the social hierarchy. The two laws of anthroposociology tell us in substance that we will find the *Homo Europaeus* or *Aryan* at the apex of both hierarchies.[48]

Lapouge mentions finally a variant on the laws of distribution of wealth and stratification: "the law of the intellectual classes." This law does not have recourse to the cephalic index, but to a measure that had notable success prior to the research of Lapouge and Ammon and that was subsequently abandoned. The law of the intellectual classes was formulated by posing a strict correlation between the dimension of the cranium and the intellectual profession.[49] In this case, too, Lapouge did not consider that the greater dimensions of the head are an acquired characteristic; he considered instead that this is a biological fact that is reproduced through heredity in the circles of the intellectual elite:

> The greater size of the head among the intellectual classes is not due to an expansion of the brain by its training and activity, the difference is, according to my own researches, quite as marked between young children of the intellectual classes as between adult members of the same groups. (Lapouge 1897: 90)

Thus, the essential data of the anthroposociological theories are always the hereditary nature of intellectual, social, and economic superiority. Environment, education, and in general the social institutions are factors that are irrelevant in this concept frame. Ammon, Lapouge, and Closson continuously reiterate that "nurture" succumbs to the laws of "nature" (in particular, that of biological heredity). The individuals found on the lower levels of the social scale must not attribute their status to the institutions that underpin society, but exclusively to their insufficient genetic gifts.[50]

Social organization, however, maintains an important role from the eugenic point of view, in that it can ease or hinder the perfection of the race. As we have already said, this is the principal obsession of the anthroposociologists. The model of society they appeal to in order to avoid this danger is expressly anti-democratic and illiberal. Ammon, no less than Lapouge, states that the organization of the hierarchy has to crystallize into separate "social classes" in such a way that marriage and reproduction of the species takes place between individuals of the same status. "Nature," writes the German anthropologist, "commends the union of like and like" (Ammon 1896: 235). The principal cause of the deterioration of the race is "panmixia," or the "crossing" of individuals with different genetic gifts. A stable distinction between social classes—"the most remarkable [institution] which the evolution of social life has produced"—instead reduces such a danger, "and thereby makes possible a much more frequent production of highly endowed individuals" (1896: 233–234).

Each attempt to modify the social stratification, improving the conditions of the more poverty-stricken classes, would in fact have deleterious effects; it would stimulate panmixia, lethal to the positive destiny of the race. "Ammon used the opinions and language of Weismann and Galton to support his claim that the various social classes represented a necessary form of natural selection, and should be preserved intact at all costs" (Weiss 1987: 95). According to the social anthropologists, egalitarian ideas[51] in fact represented the greatest peril for the human race, destined to precipitate into a nightmare of mediocrity:

> If social distinctions could be abolished and if men ceased to marry preferably within their class, the consequence would be a marked diminution in successive generations in the proportion of individuals of high endowment. There would be correspondingly a decrease in the proportion of those of exceptionally weak capacity, and the result would be even greater proportion than at present of persons of mediocre or average ability. (Ammon 1896: 236–237)

The model of representation of the social hierarchies furnished by the anthroposociologists, cleansed of its racial content, furnishes important suggestions for economists and statisticians. It is in fact one of the earliest models in which emphasis is placed on a "personal" distribution of income and wealth. It is not by chance that Vilfredo Pareto, an author who is traditionally remembered as being the first to make that step from "functional" distribution to "personal" distribution, was strongly influenced by the anthropological literature of the end of the century, even if he distanced himself from it.

IV

A Sympathetic Critic: Vilfredo Pareto

THE RECEPTION TO the anthroposociological theories given by Vilfredo Pareto is a question of a certain importance. He is probably alone among economists in giving ample space to the discussion of the theories of Ammon and Lapouge, albeit in a mix of dissent and approbation. Pareto will never come to explicitly take up "racial" and "eugenic" positions; however, his social anthropology is strongly indebted to the concepts formulated by the founders of the discipline. The concepts of "social heterogeneity," "social selection," and "circulation of the elite" to which he has frequent recourse in his work in fact owe much to Ammon and Lapouge, as he himself, in fact, recognized on more than one occasion. Pareto's theory is widely known; therefore, discussion here will be limited to those aspects that in various ways recall the anthroposociological theories.

The idea of social heterogeneity constitutes one of the pillars of the Paretian concept of social hierarchies (Spengler 1944). After having formulated his famous "income curve," Pareto explained the constancy of such curve in time, using a minimum of economic arguments. To understand the reasons underlying the unequal division of wealth, which shows up in a universal form in every type of society, economic theory has to give way to anthropology or to "social physiology," as the last chapter of the *Cours* calls it. The fact that "intelligence," "ability," "aptitude," or, more generally, "the psychical and physiological" qualities of individuals are distributed in such a way that "some possess . . . in a more eminent way than others," is due to social heterogeneity. This creates the hierarchies of income and wealth typical of all human societies, which are not due to institutional causes, like "education received" or the "social condition" of the family (Pareto 1896–1897, II: 390).

Pareto expressly declared that he was inspired, in expounding the "doctrine of social heterogeneity," by the writings of "Ammon, Lapouge and other anthropologists" (Pareto 1896: 443). He, however, distanced himself from the racist implications of those theories. The author of the *Cours* believed that the factual data on which the doctrine of the hierarchies of race was constructed were still insufficient:

> Nothing authorises us to consider the form of the cranium or the colour of the hair or eyes as exclusive characteristics, for differentiating the human races. For many anthropologists the existence of a race of blonde dolichocephalics much superior to the darker brachycephalic races, which are qualified as "inferior races," is an article of faith. Facts which are numerically scant, often badly observed and forced correlations are adduced as proof. (Pareto 1896–1897, II: 396).

The merit attributed to Ammon and Lapouge is that of having banished the "egalitarian" prejudice and of having brought the irreducible diversity of individuals to the center of attention. Their error lay in wishing to reduce that heterogeneity solely to "race" and to having understood the ethnic element as the only factor in historical explanation.[52] The "doctrine of social heterogeneity" was therefore correct when it stated that there are "innate" differences between individuals from which the unequal distribution of wealth arises. But it was in error when it stated that these differences derive from the "race," "class," or "caste" the individual belongs to.[53]

The concept of "social selection" was also discussed by Pareto in comparison with Ammon and Lapouge. For Pareto, "selection" had a "double scope"; on the one hand, it places the individual in an appropriate position in the social hierarchy; on the other, it eliminates those inept and incapable subjects that could undermine the survival of the social aggregate (Pareto 1901–1902: 541). As concerns the first goal, Pareto held that it could be reached by allowing the laws of the market and competition to act. Reaching the second goal of selection is more complex: "There are individuals who can be decidedly deleterious and dangerous for society"; selection should impede "their being reproduced in their descendants" (Pareto 1901–1902: 540–542). It would be dangerous if the "human race" were not subject to the law of "selection"; it could not then save itself from "decline" (Pareto 1906: 312). The problem is naturally "how" to eliminate elements that are held to be "inferior."

Like Ammon and Lapouge, Pareto underlined the contrast between natural selection and social selection. The first operates in a "direct" fashion, eliminating the "inferior elements" and impeding "their being reproduced in their descendants" (Pareto 1901–1902: 542). In human societies, this form of selection is hindered by social institutions that allow individuals destined by nature not to survive to do so. In Pareto's eyes, reforms of a "social" character inspired by humanitarianism and by sentimentalism stop selection from carrying out its proper function. From this perspective, there is a total communion of opinion with Ammon and Lapouge. The humanitarians and reformers, Pareto writes, "desperately" pursue the aim of "the betterment of individuals of inferior quality," not realizing that "every hope held in this regard has always been deluded" (Pareto 1896–1897, II: 552). Their error arises from not wanting to

> admit that in the human species, as in all the other living species, individuals are not born equal, they have different character, and certain individuals are adapted to the environment in which they live, others no. It is easy to believe what one wants. Humanitarians do not study the real world as it is, they create an imaginary world, one which their sentiments desire.[54]

His condemnation of the institutions that hinder selection does not, however, signify that Pareto was willing to accept the model of society proposed by Ammon and Lapouge. On the contrary, he expressed his regret that "a distinguished soul like Lapouge" had come to imagine such radical solutions for carrying out the "selection of the race" as the physical elimination of the inferior elements and the imposition of sexual duty and reproduction only on the "eugenics." This prospect appeared to Pareto to be totally "repugnant" (Pareto 1896–1897, II: 394).

He showed his approval instead for the "indirect" methods of suppression of the "inferior individuals": the penal system of reclusion and suppression of the perilous individuals; the different mortality rates and birth rates of the different social classes that contribute to suppressing "in a great number the majority of weak and malformed individuals"; the habits of vice that attract degenerate individuals, such as alcoholism, that accelerate their "degeneration" and that of their descendants. These are "means" that are still "very imperfect" but that nonetheless contribute in an indirect fashion to reaching the goal of selection; that is, the suppression of the unfit.

A further use of the concept of social selection was made by Pareto in discussing the problem of the form of the income curve. In contrast to Ammon, the author of the *Cours* did not consider that the income curve presents analogies with the normal or error curve. He did not even believe that these could be generated by chance, as Ammon implicitly supposed with his statistic simulation based on the throw of the dice. Pareto thought, on the contrary, that the income curve is much more asymmetric and compressed toward the lower part than the normal one.[55] The experiment simulated by Ammon through the casting of dice needed to be rectified, attributing a lesser probability to the dice face with the higher value, and vice versa. In a letter to Ammon in 1900, Pareto wrote:

> In the II volume of my *Cours d'èconomie politique* . . . I was able to demonstrate that the distribution of income does not depend solely on chance, but presupposes a certain law of heterogeneity among individuals.
>
> As you will see this leaves your conclusions unchanged. All that is necessary is to substitute a uniform distribution of heterogeneity with a distribution according to a given law. In other words, the dice of which you speak are not simple cubes with faces 1-2-3-4-5-6. They are instead polyhedrons having, for example, three faces marked 1, two faces marked 2, one marked 3 etc, all according to the given law.[56]

According to Pareto, this device would have permitted the formulation of a statistical law that conformed to the real trend of income distribution in a greater degree, in particular to a distribution that was highly asymmetric, flattened in the lower part of the curve. The fact that it did not coincide with the distribution of the abilities, usually normal and symmetric, was due precisely to the action of social selection.

According to Pareto, there was a lower limit beyond which it was not possible to descend that was absent from the upper part of the scale:

> Indeed, of two individuals who deviate equally from the mean of the quality, the one who has exceptional aptitudes for making money may have a very high income; but the one who has qualities equally different from the mean but in the negative direction, cannot, without dying, drop below the minimum income sufficient to sustain life. (Pareto 1906: 284)

Under the minimum income for life, the selective process operates, and eliminates the inferior elements. This explains why individuals who possess aptitudes much below the average do not appear in the income statistics—precisely due to their inability to reach subsistence level. To subsist in the lower areas of the social and economic hierarchy in substance, one needs abilities that are at least near average, since those much inferior would activate the mechanisms of selection. Pareto's use of the concept of "selection" to explain the asymmetry of the income curve is therefore in line with Ammon's;[57] there is, however, a fundamental difference concerning the idea that distribution can be considered in the same way as a casual process, that is, subject to the laws of statistical probability.[58]

Further agreement between Pareto and the anthroposociologists regards a concept that mirrors "selection" or, rather, "stability." Both concepts are taken from the work of Lapouge, in part reformulated by Pareto. As we have seen from the summary in the article by Closson, Lapouge identified "conservation and evolution" as opposing forces that hold society up. The discussion among anthroposociologists turned principally on the biological mechanisms of the transmission of the racial stock, while Pareto interpreted this process on the economic and social plane, analyzing the effects of the patrimonial heredity. "Stability," according to Pareto, contributes to "crystallizing" all the social relations, as happens in societies where there are "rigidly constituted castes"; "selection" lies instead at the origins of social dynamism. The combination of the two principles is what characterizes the "modern societies," given that here operate both "the element of stability," guaranteed "by private property and inheritance," and "the elements of change and selection" from which comes "[t]he opportunity everyone has to rise as far as he can in the social hierarchy."[59]

The different evaluation of the role of selection and of conservation emerges above all in the celebrated Paretian theory of the circulation of the elite. In this case, too, Pareto drew numerous concepts from anthropological literature, as he himself admitted. To the insinuations that his theory plagarized ideas of Gaetano Mosca, the author of the *Cours* responded:

> I have taken nothing from Mosca. On the other hand I have taken a great deal, as I have clearly stated, . . . from Ammon and a little also from Lapouge. Scholars can see moreover where I dissent from them and what I have added.[60]

The declaration by Pareto is unequivocal: He recognized that he took important notions from the theories of the anthroposociologists while dissenting from them on many decisive questions. For the circulation of the elite, the author of the *Cours* owed much to the "school" of Lapouge, even if the implications that he drew go in a completely different direction.

The first fundamental element of dissent concerns the identification of the "chosen" individuals, a term that echoes the concept of "chosen" race from Gobineau and the anthroposociologists, but that Pareto expunged of its racial significance. Pareto wrote that "by 'chosen subjects' we simply intend individuals whose activity in life is more intense" (Pareto 1896–1897, II: 416). Therefore, until there are more reliable data,[61] the idea that the "chosen subjects" posses specific anthropological characteristics must be abandoned:

> Ammon and Lapouge specify too much when they wish to give us the anthropological characteristics of this elite, of this eugenic race, identifying it with the blonde dolichocephalics. For now, this point remains obscure, and long study is still necessary, before we can establish if the psychic

> capabilities of the elite are matched by exterior, anthropometrical characteristics and for us to know precisely what these characteristics are. (Pareto 1901–1902: 133)

Having denied that it is anthropological characteristics that identify the elite, Pareto specified that recognition cannot but be based on their capabilities and their actions. The operation of market forces, in particular, was considered by the author of the *Cours* to be the least imperfect mechanism that human society had discovered for "selecting" the most capable individuals. This process would certainly be more simple if one wanted to give credit to Lapouge's theories of the elect race, given that the subjects destined for high positions could easily be identified on the basis of their facial features.[62]

Radically different from the anthroposociological theories is the Paretian concept of social mobility. The idea that individuals coming from the "lower" ranks can contaminate the aristocracies was turned upside down by Pareto. If there were not this "turnover," society would be destined to perish. It is precisely the subjects coming from the lower classes that allow the social system to continually renew itself. Social selection runs counter to the idea championed by Lapouge, that the "chosen" individuals should be organized into closed castes in order to only breed individuals with the same characteristics of superiority, given that the French anthropologist gives it such importance without drawing the necessary consequences. Selection is in fact an "agent" that permits

> the chosen subjects, born in the lower social spheres, to accede to the upper echelons. The formation of castes is contrary to this selection and is therefore a powerful agent for decadence. Some modern authors, in search of novelty, have been taken with a great fancy for the Indian institution of the caste. These authors never explain how it is that an institution that is considered so excellent did not stop India from falling prey to numerous conquerors, all lacking in castes, nor how a few thousand English are enough to maintain British dominion over a country of around two hundred million inhabitants. (Pareto 1896–1897, II: 416–417)

In the Paretian concept, prominence thus is given to the lower levels of society, whose task is to nourish the process of "circulation of the aristocracies." It is therefore indubitably true that wealth is an index of "superiority" and that the social hierarchy is in a large measure the reflection of different individual abilities.[63] However, experience shows that there are often individuals among the lower classes who are more gifted than those in the upper ranks: "Anyone who has ever spent some time with working men knows that one often meets individuals among them who are far more intelligent than this or that scientist laden with academic titles." And as Pareto punctiliously points out, this fact considerably reduces the dependability of the research carried out by Galton and the anthroposociologists on the biological hereditability of intellectual gifts (Pareto 1896–1897, II: 396).

This brief summary of Pareto's distribution theory and his conception of the social hierarchy is intended to show the intellectual debt he owed to Ammon and Lapouge. There were, naturally, notable differences—starting with the fact that he apparently did not accept that innate difference must be racial in origin, nor that the cephalic index could ascertain innate differences—differences that are, however, accompanied by a clearly declared underlying appreciation (in the letter to Ammon quoted above, for example, Pareto concludes with the wish that the theories of the German anthropologist might be "taught in all the universities"). Above all, there is agreement on the idea that individuals are heterogeneous—in contrast with the vision of homogeneity that animated classic economic thought—and that social and economic inequality are consequences of innate intellectual differences.

V

The Growth of Criticisms: Social Anthropology as a "Pseudoscience"

Pareto's case is perhaps emblematic of the entire parabola of social anthropology. As we have seen, in his works published at the turn of the century, the thinking of Ammon and Lapouge was repeatedly debated, both in the positive and the negative sense. In his *Trattato di sociologia* (1916), on the other hand, we find no further reference to the anthroposociological literature. The heterogeneous nature of society and the circulation of the elite remain the cardinal points of his social theory, but they are no longer discussed in relation to Ammon and Lapouge, while social selection disappears from the list of questions to be analyzed in the *Trattato*.

It is difficult to give an explanation for this change of attitude on Pareto's part. It is, however, a fact that, after having enjoyed great success in economic literature at the end of the 19th century, anthroposociology petered out with the coming of the new century. The debate in the economic reviews at the turn of the century shows the growing idea that anthroposociology was a "pseudo" science.

In a lengthy essay published in the *Quarterly Journal of Economics*, John Cummings (1900) attacked the primary principle from which the anthroposociologists started, that is, that the "ethnic factor" was decisive for the understanding of social evolution. Cummings held that terms like "race," "ethnic factor," or "cephalic index" were without any scientific foundation, in that Ammon and Lapouge were never able to give plausible definitions of the same (Cummings 1900: 201–202). He, moreover, denied that ability and intellectual qualities solely have a biological basis, judging the correlation with the cephalic index to be indemonstrable: "The association of index with ethical character is a matter of chance, and there is no correlation of cause and effect" (1900: 197). Cummings believed instead that it was necessary to revaluate the role of social environment, essential to the formation of the human personality, reversing the social and political philosophy of the anthroposociologists: "Environment is the matrix and to conceive man apart from environmental influences is as impossible as to conceive a cast without a mould" (1900: 199). He concluded with the wish that this perverse deviation from Social Darwinism be swept from the horizons of human knowledge and that the social sciences return to their normal path:

> Anthropologists do not present any data to justify the assumption that the cephalic index carries any mental attribute or any character with it; nor can any such contention be maintained in the face of modern psychology, which more and more is coming to regard the head-form as irrelevant to mental capacity or character, certainly where the size of the brain is disregarded. Phrenology, like astrology, has had its day; and the sort of racial phrenology with which modern anthropologists are engaged is bound to go to the same limbo. (Cummings 1900: 211)

Closson's reply, aimed at vindicating the empirical and statistical foundations of anthroposociology, was to no avail.[64] After Cummings's intervention "the school of so-called 'anthropo-sociologists' " began to be so designated in a derogatory sense.[65] The proof of this lies in the comments made by William Z. Ripley, an economist who did not lack in racist convictions, but who resolutely distanced himself from Ammon, Closson, and Lapouge.[66] The classification of the European races as used by Ripley in his works is similar to that used by the anthroposociologists. It is therefore hardly surprising that Closson at first had included him among the potential disciples of the school of Lapouge,[67] albeit an atypical one. Ripley could therefore not have completely supported Cummings's sweeping criticism. This explains his ambivalent attitude, on the one hand aimed at vindicating the legitimacy of the model of classification of the European races that he himself used, and, on the other, in line with the climate of scientific discredit for the "anthropo-sociologists." He held that the accusation launched by Cummings against the school of Lapouge was irreproachable, but that it should not involve the classification of the European races. The existence of three "physical or racial, types into which the population of Europe may be resolved" was no invention, Ripley wrote, "of an anthropo-sociologist." It dated back to the research by Broca and since then had been borne out by authoritative anthropologists (1900: 428). This did not, however, justify the implications that Lapouge, Ammon, and Closson wished to draw from the classification, which were absolutely lacking in any scientific foundation. In substance, Ripley distanced himself from the school of Lapouge, vindicating an approach still based on racial data but far from the extremes desired by anthroposociology:

> The "anthropo-sociologists" make the gratuitous assumption that certain mental traits always attach themselves to the same physical ones. I have striven manfully to disprove that any such connection exists, and that a large part of the mental characteristics of the population of Europe are attributable to physical and social environment, and not to race at all. (Ripley 1900: 427)

The distance taken by Ripley from anthroposociological thought is very important. If Closson "only dabbled briefly at the edges of American social science," Ripley, "on the other hand, was much closer to the center of the stream" (Stocking 1968: 61). His disassociation, and the criticism by Cummings, were probably decisive for the epilogue of anthroposociology in economic literature. After 1900,

Ammon, Closson, and Lapouge became total strangers in the economic arena.[68] But how so much of their theories found their way into economics journals at the turn of the century and then suddenly vanished remains puzzling (Darity 1997: 242).

A plausible explanation is that something similar to what had happened in the sociological field occurred. In this context as well, Ammon and Lapouge had a sudden popularity in the last few years of the century, even leading Emile Durkheim to introduce a specific subsection named "Anthropo-sociologie" into his journal *Année sociologique*.[69] After the devastating criticism of their theories by the anthropologist Léon Manouvrier, who qualified social anthropology as a "pseudo-science," however, Durkheim seized the chance to suppress the column, judging their ideas as "too suspect." The fortunes of social anthropology in economic literature could therefore have had a parabola not dissimilar to that seen in the sociological field. But it seems unlikely that Cummings's "intervention" was sufficient by itself to provoke the eclipse of Ammon's and Lapogue's theories.

VI

From "Nature" to "Nurture": Thorstein Veblen

ANOTHER CENTRAL QUESTION is the role of Thorstein Veblen, who was managing editor of the *Journal of Political Economics* between 1896 and 1906. How is the appearance and subsequent abrupt disappearance of the anthroposociological theories related to his interest in anthropology, evolution, and biology?[70] From the anthropological point of view, the biographies regarding Veblen tend to highlight the ascendancy exercised over him by Franz Boas, with whom Veblen came into contact when he arrived in Chicago, where he also had the opportunity to know Carlos Closson. Only Spindler (2002: 23), however, mentions Closson as one of the authors who may have influenced the anthropological thought of the American institutionalist.

As we know, Veblen was very attracted to biological explanations of social and economic phenomena: His famous "Why Economics Is Not an Evolutionary Science" (1898: 373) begins with an approving quote of Lapouge's peremptory statement (1897: 54): "Anthropology is destined to revolutionize the political and the social sciences as radically as bacteriology has revolutionised the science of medicine" (where Lapogue is depicted as an "eminent anthropologist"). In Veblen's work there is also a recurring use of the concept of race and there are several remarks about the characteristic head shapes of European races. From this point of view, there is little doubt that Lapouge, Ammon, and Closson (as well as other racist physical anthropologists) were the source of Veblen's anthropological thought. It is, however, difficult to assess the real impact that anthroposociologists had on the founder of American institutionalism, and a full account of Veblen's debt to Ammon, Closson, and Lapouge has yet to be published.

Their works are quoted in diverse circumstances (see Edgell 2001: 60), and the idea that the cephalic index was fundamental in the identification of the racial types appears to be substantially agreed upon. One of the most explicit citations is to be found in the *Theory of Business Enterprise*, where Veblen states that races possess not only physical differences but also specific intellectual, psychological, and even religious characteristics. The American institutionalist writes:

> If the researches of such students as Ammon, Ripley, Lapouge, Closson, and others that might be named are taken at their face value, it appears that the towns differ perceptibly from the open country in point of race; and that the migration from the country into the industrial towns has a selective effect of such a kind that a larger proportion of one racial stock than of another resorts to the towns. The towns, in those countries where data are available, show a larger admixture of the dolicho-blond stock than the open country. This seems to argue that the dolicho-blond stock, or the racial mixture of the towns in which there is a relatively large admixture of the dolichoblond, is perceptibly more efficient in the machine industries, more readily inclined to think in materialistic terms, more given to radical innovation, less bound by convention and prescription. This generalization is strengthened by the fact that the more dolicho-blond regions are also, on the whole, more socialistic than those in which this element is less in evidence. At the same time they are industrially in advance of the latter in the matter of machine industry; and they are also Protestant (irreligious) rather than Catholic. (Veblen 1904: 350, n. 24)

These sentences demonstrate the undoubtedly strong influence exercised by the anthroposociologists, which was reinforced by the fact that Veblen believed in the hereditary nature of cultural attitudes and inclinations that characterized diverse ethnic and racial groups. It is in this sense that Tilman (1996: 57) defines Veblen as "a racialist." Some scholars see this cultural ascendancy as one of the major limits of Veblen's thought. Abram L. Harris, for instance, in his extensive investigation of Veblen's work, wrote that "the anthropology from which Veblen derived his theory of three European racial types—that differ in both physical and mental characteristics—has long been discarded" (Harris 1951: 66).

On the other hand, it is equally certain that Veblen largely attenuated the biological and racial determinism of the anthroposociologists. He expressed notable skepticism about the possibility of identifying pure races[71] and, more generally, about the fact that a certain type of race might be genetically superior. For Veblen, the diverse attitudes or spiritual characteristics were to be attributed to the diverse social and cultural contexts rather than to racial differences. In 1914, in fact, he wrote:

> It is perhaps as needless to insist on this spiritual difference between the various racial stocks as it would be difficult to determine the specific differences that are known to exist, or to exhibit them convincingly in detail. To some such ground much of the distinctive character of different peoples is no doubt to be assigned, though much also may as well be traceable to local peculiarities of environment and of institutional circumstances. (Veblen [1914] 1922: 111)

A further element that differentiates him from the anthroposociologists is his insistence on the hybrid nature of the races—in particular the European and Japanese ones—an aspect that further weakened the theory of the preeminence of one race over another:

> It should be noted in the same connection that hybrid peoples, such as those of Europe or of Japan, where somewhat widely distinct racial stocks are mingled, should afford a great variety and wide individual variation of native gifts, in workmanship as in other respects. Hybrid stocks, indeed, have a wider range of usual variability than the combined extreme limits of the racial types that enter into the composition of the hybrid.[72]

It is certain, therefore, that Veblen was fascinated by the work of Ammon, Closson, and Lapouge. But, like Ripley, he abandoned the biological determinism typical of the anthroposociologists, shifting his attention to institutional conditioning (i.e., nurture) and realizing the substantially hybrid nature of the European races, all of which makes it difficult for us to say whether he was more or less decisive in anthroposociology's fate.

VII

Conclusion: The Epilogue of Social Anthropology

The reasons behind the incredible rise of anthroposociology in economic literature are probably to be sought in the general cultural context of the 19th century. As Jennifer M. Hecht (2000: 304) wrote, "we have little idea today of how utterly convinced many people were that the European races were physiologically measurable and socially irreconcilable." The fact that, with the beginning of the new century, both economics and sociology rejected the anthroposociology of Ammon and Lapouge appears rather due to the excessive radicalism of their theories than to their racism. It would be wrong, therefore, "to conclude that . . . that race had ceased to be an explanatory variable" (Llobera 2003: 118). Race-based (and eugenic) social science more generally was pervasive outside the boundaries of anthroposociology, and survived it. The economic arena probably ceased to be a sounding board for Ammon's and Lapouge's theories because "they were too extreme, and more palatable (but still racist) alternatives won out."[73] At least until the 1920s, racist ideas continued to be promulgated (above all, in the debate on immigration) and some economists maintained a "genetic racial position," but this does not seem to be directly linked to the anthroposociological theories we have discussed.[74]

This does not mean, however, that anthroposociology failed to reach its goal. By that point, the way had been opened. Even without the showplace of the French- and English-speaking socioeconomic journals, anthroposociology was able to consolidate its position due mainly to the *Politisch-Anthropologische Revue*, under the editorship of Ammon and Ludwing Woltmann. In 1909, Lapouge himself admitted that anthroposociology had become a science that was exclusively "German"; the eugenic National Socialist project was beginning to materialize, with Ammon and Lapouge as its household gods.

Notes

1. See, for example, *Social Anthropology*, the journal of the European Association of Social Anthropology, founded in 1989.

2. See Cherry (1976); Levy and Peart (2004); Peart and Levy (2003); Colander, Prasch, and Sheth (2004); and the recent symposiums of *History of Political Economy* ("Prejudice and the History of Economics: A Minisymposium") and *American Journal of Economics and Sociology* ("Symposium on Eugenics During the American Progressive Era"), with essays by Weintraub (2003), Leonard (2003), Bateman (2003), Levy and Peart (2003), Coleman (2003), Leonard (2005c), Cot (2005), Dimand (2005), Peart and Levy (2005b), Rutherford (2005), and Smith (2005).

3. Mucchielli (1997); Hecht (1999); Llobera (2003).

4. A brief mention of the thought of the anthroposociologists—in particular, their concept of "social selection"—can be found in the book by G. Hodgson (2004), *The Evolution of Institutional Economics.*

5. "This new school, or new science . . . , has been called by its creators Anthropo-sociology, or Social Anthropology, and it is already promulgated by numerous champions, among whom Lapouge in France and Ammon in Germany occupy positions of special prominence" (Loria 1899: 283).

6. For biographical detail on Lapouge, I have drawn upon Ackerknecht (1950: 287–289); Hecht (2000: 293–294); Llobera (2003: 120–121); Quinlan (1998: 394–395); Weiss (1987: 93–94).

7. Ammon to Lapouge, February 4, 1893, cited in Hecht (2000: 295).

8. For Ammon, as for Lapouge, "the laws of nature formulated by Darwin are absolute in the most stringent meaning of the term" (Stark 1961: 50).

9. On the "Social Darwinism" of the two founders of anthroposociology, see Stark (1961). On Lapouge in particular, see Hecht (1999), Bèjiin (1982), Boissel (1982), and Clark (1981); on Ammon, see Gasman (1971) and Stein (1988).

10. What distinguishes the authors quoted is their judgment on the outcome of such a process, that is, whether the creation of inequality has been a positive factor for civilization, but not the original concept of the equality of the human being.

11. Harris (1968: 109). The question concerns all the social sciences in general, not only anthropology. As far as political economy is concerned, in particular, see Levy and Peart (2003, 2004, 2005a). The two scholars have drawn attention precisely to the shift that is recorded in the economic field from the paradigm of social homogeneity, which characterized the classical tradition, to a concept "involving racial differences." "In the second half of the 19th century, theories of racial heterogeneity were much discussed in British anthropological circles, and attacks on equal competence emerged from within economics itself" (2004: 125).

12. Gobineau would not seem to have had a decisive influence on the initial formation of Lapouge. According to Ammon, the *Essay sur l'inegalité des races humaines* was not known to Lapouge until 1894, if not through Paul Broca (Ammon 1898). Later, Gobineau would be regarded as a spiritual father by anthroposociology. See Lapouge (1897: 56): "At the origin of every discovery there is usually one man of genius. The originating genius of anthropo-sociology was the Count de Gobineau. . . . Gobineau was no scientist, but a wide traveller and a man of erudition and reflection. His fundamental idea was the superiority of the blond race. . . . Anthropo-sociology dates from the *Essay sur l'inegalité*."

13. "Gobineau's *Essai*, which was published before Darwin's *Origin of Species*, rested primarily on second-hand historical and linguistic 'evidence'; the French aristocrat never attempted to incorporate biological or anthropological theories into his philosophy of history" (Weiss 1987: 93–94).

14. Calculated comparing the width and length of the head. The notion of "cephalic index" was connected to Lavater and Gall's "phrenology," even if its introduction around 1845 by the Swedish scientist André Retzius was initially critical of phrenology (see Poliakov 1971: 299).

15. There is instead no precise psychological characterization of *Homo Mediterraneus*.

16. As Closson wrote (1900a: 399), "Lapouge has measured some 12,000 subjects and Ammon 22,962, subsequently analysing from various points of view the data thus obtained."

17. "Measurement was essential to racial science; measurement offered a tool for determining racial differences and a potential means for ranking the separate races. Before the advent of intelligence testing . . . racial scientists relied on anthropometric measurement, especially skull measurement, or craniometry" (Leonard 2003: 689).

18. "Nature is all that a man brings with himself into the world; nurture is every influence from without that effects him after his birth" (qtd. in Field 1911: 14). The fundamental idea of eugenics, a discipline founded by Galton, is that the mental qualities, no less than the physical ones, are subject to the principles of biological heredity (Field 1911: 5–6). The theory was formulated by Galton for the first time in some writings around the mid-1860s and then presented in systematic form in the volume *Hereditary Genius* (1869), in which abundant genealogical material concerning the history of the families of men of genius is collected (on Galton, see Cowan 1972).

19. According to Lapouge, the Enlightenment's triad of "Liberty, Equality, and Fraternity" should have been replaced with a motto that paid homage to the Darwinian revolution: "Determinism, Inequality, and Selection."

20. Lapouge (1896: 198). In 1909, Lapouge would define anthroposociology as the science "which has as its object the study of the reciprocal action of race and environment" (1909: vii).

21. On this question, Stark has identified a partial difference between Ammon and Lapouge. In the former, an optimistic vision of social selection seems to prevail, generally holding it to be capable of generating a progressive process. Lapouge, on the contrary, for motives that we will return to later, appears much more obsessed with the idea of an involutive movement, in which social selection operates to the detriment of the race (see Stark 1961: 49–55).

22. As, for example, racial contamination, to which we owe the reduction in fertility in the pure dolichocephalics; or celibacy, a strategy widespread among the members of the upper classes in order to maintain a high socioeconomic status.

23. Battini (1995: 199). An example of a historical type was offered by the French Revolution: "The failure of the Revolution was a clamorous one . . . this was above all the substitution of the brachycephalic in place of the blonde dolicho in holding power. . . . Through the Revolution the brachycephalic conquered power, and with democratic evolution this power tends to be concentrated in the hands of the lower classes, the most brachycephalic. The Aryan as I have defined him is quite another thing, he is the *Homo Europaeus*, a race which made France great and which is today almost extinct here" (Lapouge 1899a: VII; 22 and 464).

24. Giove (2001: 127). Among the means for such an objective there was also artificial insemination, to be imposed only on eugenic individuals.

25. Lapouge (1896: 471); Battini (1995: 201).

26. Lapouge (1896: 480ff); Battini (1995: 202).

27. Hecht (1999: 3). See also Stein (1988: 57–58): "Whether or not physical anthropology is considered a true science, there is little doubt that the anthropologists who discovered all the measurable divergent physical, psychological and mental characteristics of the various races thought they were scientific. And so did the general public."

28. From the mid-1890s, Lapouge began to publish "in new social science journals, such as René Worms' *Revue international de sociologie*, Charles Gide's *Revue d'economie politique* and Emile Durkheim's *Année sociologique*" (Schneider 1990: 62).

29. "During the late nineteenth and early twentieth centuries, the pages of the *Journal of Political Economy* and the *Quarterly Journal of Economics* were filled with studies by anthropometrics ranking the relative qualities of racial groups on both physical and mental dimensions. The distinctive nineteenth-century concept of race blended nationality, culture and genetics to distinguish human groups. Craniometry was a popular subject matter in *economics* journal of the day" (Darity 1997: 242).

30. A brief reference to Closson's university years may be found in the autobiography by Robert Morss Lovett, who attended Harvard at the same time and himself taught English literature at Chicago. See Lovett (1948: 39–42).

31. In the Annual Register of University of Chicago, Closson is listed as faculty for the school years 1894–1895 and 1895–1896 only (p. 19). At this time, Thorstein Veblen is listed as "Tutor in Political Economy." I would like to thank David Pavelich, Reference and Instruction Librarian of the University of Chicago, for this information.

32. See Stocking (1968: 60): "From 1895 to 1900 anthroposociological doctrines were summarized and explicated for American social scientists by Closson, who in the interval between his undergraduate years at Harvard and his later life as a real estate broker on the West Coast was an instructor of economics at the University of Chicago."

33. Closson also speaks at times of "statistical anthropology" to underline the quantitative approach that marks the new discipline. The essential aspect remains, however, the overcoming of the traditional boundaries between disciplines produced in the field of social sciences by the advent of the new "school": "The work of Ammon himself, together with that of De Lapouge, has not only brought statistical anthropology into close relation with politics, economics, ethics, psychology, the interpretation of history and especially sociology; but conversely it has also transformed the methods of anthropological investigation itself, enriching that science with new categories and distinctions, and with a multitude of new problems" (Closson 1896c: 411–412).

34. Closson (1896b: 93–94). This is the so-called Law of Social Stratification, to which we will return.

35. "Law of Displacement. The dolichocephalic type, being more enterprising and restless, is more largely represented among the migrants to the cities and the foreign countries. The law affirms the greater mobility of *Homo Europaeus*" (Closson 1896b: 94). In Lapouge (1897: 416–418), the "Law of Displacement" is articulated into diverse subspecies: (a) Law of Altitudes: The European man settles chiefly in the plains; (b) Law of Distribution of the cities: The most important cities are all located where there is a prevalence of dolichocephalics; (c) Law of urban Indexes: The cephalic index of the urban populations is lower than that of the rural populations; (d) Law of Mobility (or Ammon's Law, to whom the discovery that the urban populations are largely dolichocephalic is attributed): The dolichocephalic populations are characterized by greater mobility.

36. The spread of the theories of anthroposociology in the English-speaking press, as ever through the work of Closson, extended to the *American Journal of Sociology*. Besides publishing his own writings on the hierarchy of the European races (Closson 1897), the economist translated the introduction to one of Lapouge's main books, *L'aryen*, in which the French anthropologist gives his reasons for the choice of the term to define the "elect" dolichocephalic race (see Lapouge 1899a, 1899b).

37. Closson discloses in the initial notes that he has the translation of the entire book underway, but we have no news that this translation was ever published.

38. "Known at the time as the 'law of error,' the formulation derived from the analysis by the German mathematician Carl Friedrich Gauss of errors made in the measurement of 'true' physical quantities—for example, planetary position in astronomy. Portrayed graphically, the Gaussian distribution formed the now familiar bell curve" (Kevles 1995: 13).

39. The formula concerned is $y = Ye^{-h^2x^2}$. "In the formula x signifies the degree of variation from the mean, y the relative frequency of the appearance of this variation (that is its 'probability'), Y the frequency of the mean, e the basis of the hyperbolic logarithms, h the so-called coefficient of precision which determines whether the frequency of occurrence shall diminish more or less rapidly with the increasing degree of variation from the mean. The quantity e is a constant, fixed once and for all; Y and h are constants which may be altered according as the formula is differently utilized" (Ammon 1896: 214).

40. The English scientist, in his studies on genius, had tried to demonstrate the convergence between the distribution of aptitude and social status. His reasoning, however, stopped at the idea of a correspondence between what he called "reputation" and "ability": High reputation, Galton wrote, "is a pretty accurate test of high ability" (Galton 1869: 2).

41. "The highest possible throw is that in which the sum of the spot is equal to 24, and this can occur only in a single way, namely, that every die shows six spots. In our parable this suggests that among 1296 individuals will be found only a single one in whom the mental, moral economic, and bodily traits all attain the highest grade" (Ammon 1896: 207).

42. The "throw showing 1 spot on each of the four dice, signifies in our comparison a man exceptionally poor in all four groups of traits, weak-winded and sickly" (1896: 208–209).

43. The imperfect correspondence between the two curves is due to the fact that Ammon's experiment is conducted imagining "the combination of only a few elements or traits; the more elements co-operate the more traits must meet in a genius, the broader is the curve in the centre, and the more tapering is its attenuation both above and below" (1896: 218).

44. These individuals "are so poorly endowed that they can only with difficulty and in an incomplete way take part in the life of society. In this category belong not only those who are not able to accomplish work which requires intelligence and skill, but also those of insufficient moral character" (Ammon 1896: 218).

45. Ammon took "the distribution of incomes from the income-tax statistics of Saxony for the year 1890" (Staehle 1943: 77).

46. "Laws of the distribution of wealth: In a country jointly inhabited by *Homo Europeaus*, the former element possesses more than its proportionate share of wealth" (Lapouge 1897: 61).

47. "The cephalic index is lower and the proportion of dolichocephalic greater among the higher classes than among the lower classes in each community "(Lapouge 1897: 87). See also Closson's formulation: "Law of Social Stratification. The dolichocephalic type, being psychologically more domineering and ambitious, is more generally represented among the higher and ruling classes; the brachycephalic type is found generally in a subordinate social position. The law, then, affirms the social superiority of *Homo Europaeus*" (Closson 1896b: 93–94).

48. For Lapouge, "the 'Aryan race' is the only race capable of high social, intellectual, and cultural achievements, and is in fact the true biological underpinning of Western civilization" (Weiss 1987: 93–94).

49. "*Among intellectual workers the absolute dimension of the head and particularly the breadth, are greater than the average*" (Lapouge 1897: 90, original emphasis; in the following, the emphasis, if not otherwise stated, is in the original text). The French anthropologist clarifies that the law of the intellectual classes is different from the law of distribution and stratification, and can at times be in conflict with them: "The intellectual worker tends to possess *an exceptional breadth of head more marked than the exceptional length*. He is likely then to be *less dolichocephalic* than the average of his social compeers; he is likely to possess a degree of eurycephaly which may be easily confused with brachycephaly" (1897: 91).

50. "What is lacking here is . . . insufficient strength in the intellectual, moral, and economic traits to advance the individual to a higher position" (Ammon 1896: 230).

51. The democratic and egalitarian aspiration is in Lapouge's opinion an intellectual perversion inspired "by the dreams of that most visionary of all centuries, the eighteenth" (Lapouge 1896: 259).

52. In fact, Pareto observes that for some "authors, like Lapouge, the zoological race of men explains everything. This is one of the reasons why he was overlooked. It is his merit to have placed it well in evidence, even if he exaggerated its action, but considering it as unique it is nothing less than pure fiction" (Pareto 1898: 158).

53. "To say that there are in society men who have certain qualities in a more eminent measure than others and to say that there exists a class of men who are in the absolute sense *better* is already not the same thing" (Pareto 1896–1897, II: 392).

54. Pareto (1901–1902: 554). See also Pareto (1906: 312–313): "The *humanitarians* can close their eyes and deliberately ignore this truth, but that in no way changes the fact. Some degenerate elements which have to be eliminated by selection are born in every species. The unhappiness caused by

this destruction is the price paid for perfecting the race, it is one of those many cases where the good of the individual is at variance with the good of the species."

55. At the beginning, Pareto had criticized the representation proposed by Ammon, which is, however, revaluated in the work following the *Cours*. Pareto wrote "that which is called social pyramid is in reality a kind of spinning top," that "which is indicated by Otto Ammon, which seems quite plausible" (Pareto 1901–1902: 19).

56. Vilfredo Pareto to Otto Ammon, November 30, 1900, Banca Popolare di Sondrio, Fondo Vilfredo Pareto (IT PopSo FP R11C023). I am grateful to Pier Carlo Della Ferrera and the Banca Popolare di Sondrio for permission to cite this unpublished letter of Pareto.

57. Pareto and Ammon are frequently associated for having first highlighted "that the distribution of incomes is strongly skewed" (Staehle 1943: 77; Lebergott 1959: 328).

58. "When Pareto said that his results did not depend on mere chance, he really meant that the distribution of incomes had nothing whatsoever to do with either the normal curve or any of its skewed variants" (Staehle 1943: 78).

59. Pareto (1906: 304). Despite the several implications that Pareto draws from these principles, he underlines how Closson (1896d), in the article in which he discusses the concepts of selection and stability, "has made an important contribution to this line of study" (Pareto 1897: 502).

60. V. Pareto to G. Prezzolini, December 17, 1903, in Pareto (1973, I: 507) see also Pareto (1906: 312, 1901–1902: 131) where the research of Ammon and Lapouge is mentioned as "scientific confirmation" of the fact that "the history of human society is, in the greater part, the history of the alternation of aristocracies."

61. This suspension of judgment on the theories of social anthropology, to be left until the facts are verified, recurs in diverse instances in the works of Pareto. A propos of the thesis sustained by Lapouge, for example, that in ancient Rome, the elite had declined because of "the exhaustion of eugenics, of men of the superior race," Pareto comments that "we lack facts to allow for the absolute acceptance or rejection of this opinion" (Pareto 1901–1902: 157).

62. "If, in fact, from some exterior signs, such as the shape of the skull, the colour of the hair, of the eyes etc, it were possible to recognise the character and habits of men, the problem of the best social organisation would be easily resolved. Unfortunately, these theories have still an uncertain link with reality, and for the moment we know of no other means of choosing men except that of finding out what they know how to do, placing them in competition one with the other. This takes place, even if in a most imperfect mode, in our societies, and history demonstrates that their progress has been intimately linked to the extension of this custom" (Pareto 1901–1902: 342). Pareto proposes an

interesting parallel between Lapouge's theory and that of Plato, whose "golden race" in a version by the French anthropologist "would be that of the blonde dolichocephalics."

63. "The so-called superior classes are also generally also the richest. These classes constitute an *elite*, an *aristocracy* (in the etymological sense: *αριστος* = the best)" (Pareto 1901–1902: 131).

64. "While the economist and the worker in many branches of social science can draw without labour and without price upon a vast mass of official statistics, the anthropo-sociologists have had to collect their own data by the slow process of measuring individuals and by the vastly slower process of persuading them one by one to allow themselves to be measured" (Closson 1900a: 398).

65. Although it appeared in an anthropological journal, the position taken by the Italian economist Achille Loria is also significant. Loria's interest is in Ammon's theory of the social hierarchy, and in particular in his attempt to prove the coincidence between the ability curve and the income curve. The Italian economist holds that Ammon does not prove anything from this point of view, and that no coincidence is possible "in regard to the question under discussion" (1899: 293). Loria concludes with a comprehensive negative judgment of the scientific project of the anthroposociologists, held to be deleterious to the development of the social sciences. The theory of Ammon, "with its errors, its paradoxes, and the absurdity of the practical conclusions to which it leads, constitutes a direct proof, drawn from anthropologic and biologic studies themselves, of the fallacy of a scientific tendency which pretends to turn social science into an appendage of anthropology" (Loria 1899: 296).

66. "Although Ripley made a special point of dissociating himself from the anthroposociologists," both Closson and Cummings "insisted on the linkage" (Stocking 1968: 323, n. 42).

67. "The anthropo-sociological school may well be content to welcome Professor Ripley . . . as on the whole an adherent, if indeed a somewhat eclectic one" (Closson 1899: 241).

68. To our knowledge, echoes of the anthroposociological theories are to be found only in the works of Veblen, which we will discuss later, in Sorokin (1927) and in von Mises (see his *Socialism* [1922] 1951: 314–327).

69. The French sociologist did not like anthroposociology, "but as long as Lapouge appeared to be a good scientist, Durkheim felt obliged to permit extremely positive reviews of his work to be printed in the journal" (Hecht 1999: 8). On the relations between the Durkheimiens and the anthroposociologists, see also Mucchielli (1997) and Llobera (2003: 110–115).

70. This is underlined by Darity (1997: 242), according to whom a partial explanation, with respect to the *Journal of Political Economics*, of the spread of anthroposociological theory may be the interest of Thorstein Veblen in such theories.

71. "There neither is nor ever has been a pure-bred dolicho-blond individual" (Veblen 1913b: 469; Edgell 2001: 60; see also Veblen 1913a).

72. Veblen ([1914] 1922: 111–112). According to Edell, Veblen showed himself to be quite cautious in the use of the actual term "race," preferring expressions like "social groups" or "ethnic types." "Veblen's choice of terminology reflects his view that culture rather than nature was more important in the social scientific analysis of change" (Edgell 2001: 60).

73. I'm grateful to an anonymous referee for underscoring this point.

74. On the persistence of race-thinking in economics at the beginning of the 20th century, see, among others, Cherry (1976), Leonard (2003, 2005a, 2005b, 2005c), Cot (2005), and Dimand (2005).

References

Ackerknecht, Lucy K. (1950). "Some Political Predictions of an Early Anthropologist." *American Anthropologist* 52(2): 287–291.

Ammon, Otto. (1890). *Anthropologische Untersuchungen der Wehrpflichtigen in Baden.* Hamburg: Verlagsanstalt und Druckerei Actien-Gesellschaft.

——. (1893). *Die naturliche Auslese beim Menschen: auf Grund der Ergebnisse der anthropologischen Untersuchungen der Wehrpflichtigen in Baden und anderer Materialien.* Jena: G. Fischer.

——. (1895). *Die Gesellschaftsordnung und ihre naturlichen Grundlagen: Entwurf einer Sozial-Anthropologie zum Gebrauch fur alle Gebildeten, die sich mit sozialen Fragen befassen.* Jena: Fischer [*The Social Order and its Natural Bases.* Partially trans. in Ammon 1896].

——. (1896). "Some Social Applications of the Doctrine of Probability." *Journal of Political Economy* 7(2): 204–237.

——. (1898). "Histoire d'une idée: L'anthropo-sociologie." *Revue internationale de sociologie* 6(3): 145–181.

——. (1900). *L'ordre social et ses bases naturelles: Esquisse d'une anthroposociologie.* Trans. H. Muffang. Paris: Fontemoing.

Bateman, Bradley W. (2003). "Race, Intellectual History, and American Economics: A Prolegomenon to the Past." *History of Political Economy* 35(4): 713–730.

Battini, Michele. (1995). *L'ordine della gerarchia: I contributi reazionari e progressisti alle crisi della democrazia in Francia, 1879–1914.* Turin: Bollati Boringhieri.

Bèjiin, André. (1982). "Le sang, le sens et le travail: George Vacher de Lapouge, darwinist social, fondateur de l'anthroposociologie." *Cahiers internationaux de sociologie* 73: 323–343.

Boissel, Jean. (1982). "George Vacher de Lapouge: Un socialiste révolutionnaire darwinien." *Nouvelle Ecole* 13: 59–83.

Cherry, Robert. (1976). "Racial Thought and the Early Economics Profession." *Review of Social Economy* 34: 147–163.

Clark, Linda L. (1981). "Social Darwinism in France." *Journal of Modern History* 53(1): 1039–1040.

Closson, Carlos C. (1896a). "Dissociation by Displacement: A Phase of Social Selection." *Quarterly Journal of Economics* 10(2): 156–186.

——. (1896b). "Ethnic Stratification and Displacement." *Quarterly Journal of Economics* 11(1): 92–104.

——. (1896c). "Recent Progress of Social-Anthropology." *Journal of Political Economy* 4(3): 410–412.

——. (1896d). "Social Selection." *Journal of Political Economy* 4(4): 449–466.

——. (1897). "The Hierarchy of European Races." *American Journal of Sociology* 3(3): 314–327.

——. (1898). "The Pedagogical Significance of the Cephalic Index." *Journal of Political Economy* 6(2): 254–265.

——. (1899a). "Further Data of Anthropo-sociology." *Journal of Political Economy* 7(2): 238–252.

——. (1899b). "The Races of Europe." *Journal of Political Economy* 8(1): 58–88.

——. (1900a). "A Critic of Anthropo-Sociology." *Journal of Political Economy* 8(3): 397–410.

——. (1900b). "The Real Opportunity of the So-Called Anglo-Saxon Race." *Journal of Political Economy* 9(1): 76–97.

——. (1900c). "Heredity and Environment: A Rejoinder." *Quarterly Journal of Economics* 15(1): 143–146.

Colander, David, Robert E. Prasch, and Falguni A. Sheth, eds. (2004). *Race, Liberalism, and Economics.* Ann Arbor: University of Michigan Press.

Coleman, William (2003). "Anti-Semitism in Anti-Economics." *History of Political Economy* 35(4): 759–777.

Cot, Annie L. (2005). " 'Breed Out the Unfit and Breed In the Fit'. Irving Fisher, Economics, and the Science of Heredity." *American Journal of Economics and Sociology* 64(3): 793–826.

Cowan, Ruth Schwartz. (1972). "Francis Galton's Statistical Ideas: The Influence of Eugenics." *Isis* 63(4): 509–528.

Cummings, John. (1900). "Ethnic Factors and the Movement of Population." *Quarterly Journal of Economics* 14(2): 171–211.

Darity, William A., Jr. (1997). "Soundings and Silences on Race and Social Change: Abram Harris, Jr. in the Great Depression." In *A Different Vision: African American Economic Thought*, Vol. 1. Ed. Thomas D. Boston. New York: Routledge.

Dimand, Robert W. (2005). "Economists and the Shadow of 'The Other' Before 1914." *American Journal of Economics and Sociology* 64(3): 827–850.

Dorfman, Joseph. ([1934] 1947). *Thorstein Veblen and His America.* New York: Viking Press.

Drouard, Alain. (2005). "Otto Ammon." In *Dictionnaire de la pensée sociologique.* Paris: PUF.

Edgell, Stephen. (2001). *Veblen in Perspective: His Life and Thought.* Armonk, NY: ME Sharpe.

Field, James A. (1911). "The Progress of Eugenics." *Quarterly Journal of Economics* 26(1): 1–67.

Galton, Francis. ([1869] 1914). *Hereditary Genius: An Inquiry into Its Laws and Consequences.* London: MacMillan.

Gasman, Daniel. (1971). *The Scientific Origins of National Socialism, Social Darwinism in Ernst Haeckel and the German Monist League.* London: Macdonald.

Giove, Nicoletta. (2001). *Le razze in provetta: Georges Vacher de Lapouge e l'antropologia sociale razzista.* Padua: Il Poligrafo.

Gordon, Scott. (1991). *The History and Philosophy of Social Science.* London: Routledge.

Hankins, Frank H. (1931). *The Racial Basis of Civilization,* rev. ed. New York: A. A. Knopf.

Harris, Abram L. (1951). "Veblen and the Social Phenomenon of Capitalism." *American Economic Review* 41(2): 66–77.

Harris, Marvin. (1968). *The Rise of Anthropological Theory.* New York: Thomas Y. Crowell.

Hawkins, Mike. (1997). *Social Darwinism in European and American Thought.* Cambridge: Cambridge University Press.

Hecht, Jennifer Michael. (1999). "The Solvency of Metaphysics. The Debate Over Racial Science and Moral Philosophy in France, 1890–1919." *Isis* 90(1): 1–24.

——. (2000). "Vacher de Lapouge and the Rise of Nazi Sciences." *Journal of the History of Ideas* 61(22): 285–304.

——. (2003). *The End of the Soul: Scientific Modernity, Atheism, and Anthropology in France.* New York: Columbia University Press.

Hodgson, Geoffrey M. (2004). *The Evolution of Institutional Economics: Agency, Structure and Darwinism in American Institutionalism.* London: Routledge.

Kevles, Daniel J. (1995). *In the Name of Eugenics: Genetics and the Uses of Human Heredity.* Cambridge: Cambridge University Press.

Lapouge, George Vacher de. (1896). *Les sélections sociales: Course libre de Science Politique professé a l'Université de Montpellier 1888–1889.* Paris: A. Fontemoing.

——. (1897a). "The Fundamental Laws of Anthropo-sociology." *Journal of Political Economy* 6(1): 54–92.

——. (1897b). "Corrélations financières de l'indice céphalique." *Revue d'économie politique* 11(3): 257–279.

——. (1899a). *L'aryen, son rôle social: Course libre de Science Politique professé a l'Université de Montpellier 1889–1890.* Paris: A. Fontemoing.

——. (1899b). "Old and New Aspect of the Aryan Question." *American Journal of Sociology* 5(3): 329–346.

——. (1909). *Race et milieu social. Essais d'anthroposociologie.* Paris: Rivière.

Lebergott, Stanley. (1959). "The Shape of the Income Distribution." *American Economic Review* 50(3): 528–547.

Leonard, Thomas C. (2003). " 'More Merciful and Not Less Effective': Eugenics and American Economics in the Progressive Era." *History of Political Economy* 35(4): 709–734.

——. (2005a). "Retrospectives: Eugenics and Economics in the Progressive Era." *Journal of Economic Perspectives* 19(4): 207–224.

——. (2005b). "Mistaking Eugenics for Social Darwinism: Why Eugenics Is Missing from the History of American Economics." *History of Political Economy* 37(supplement): 200–233.

——. (2005c). "Protecting Family and Race: The Progressive Case for Regulating Women's Work." *American Journal of Economics and Sociology* 64(3): 757–791.

Levy, David M., and Sandra J. Peart. (2003). " 'Who Are the Canters?' The Coalition of Evangelical-Economic Egalitarians." *History of Political Economy* 35(4): 731–757.

——. (2004). "Statistical Prejudice: From Eugenics to Immigration." *European Journal of Political Economy* 20(1): 5–22.

Llobera, Josep R. (2003). *The Making of Totalitarian Thought.* New York: Berg.

Loria, Achille. (1899). "Social Anthropology: A Review." *American Anthropologist* 1(2): 283–296.

Lovett, Robert Morss. (1948). *All Our Years, the Autobiography of Robert Morss Lovett.* New York: Viking Press.

Mises, Ludwig von. ([1922] 1951). *Socialism: An Economic and Sociological Analysis.* New Haven, CT: Yale University Press.

Mosse, G. L. (1978). *Toward the Final Solution: A History of European Racism.* New York: Haward Ferting.

Mucchielli, Laurent. (1997). "Sociologie versus anthropologie raciale: L'engagement des sociologies durkheimiens dans le contexte 'fin de siècle' (1885–1914)." *Gradhiva: Revue d'histoire et d'archives de l'anthropologie* 21: 77–95.

Pareto, Vilfredo. (1896). "La curva delle entrate e le osservazioni del prof. Edgeworth." *Giornale degli economisti* 13(10): 439–448.

——. (1896–1897). *Cours d'économie politique.* Vols. I–II. Lausanne: F. Rouge.

——. (1897). "The New Theories of Economics." *Journal of Political Economy* 5(4): 485–502.

——. (1898). "Del materialismo storico." *Zeitschrift für Sozialwissenschaft* 1. Rpt. in V. Pareto 1980. *Scritti sociologici minori*. Torino: Utet.

——. (1901–1902). *Les systèmes socialistes*, Vol. I–II. Paris: Giard & Brière.

——. (1906). *Manuale di economia politica con una introduzione alla scienza sociale*. Milan: Società Editrice Libraria.

——. (1916). *Trattato di Sociologia Generale*. Florence: Barbera.

——. (1973). *Epistolario: 1890–1923*. Ed. G. Busino. Rome: Accademia Nazionale dei Lincei.

Peart, Sandra J., and David M. Levy. (2003). "Denying Human Homogeneity: Eugenics and the Making of Post-Classical Economics." *Journal of the History of Economic Thought* 25(3): 261–288.

——. (2005a). *The "Vanity of the Philosopher": From Equality to Hierarchy in Post-Classical Economics*. Ann Arbor: University of Michigan Press.

——. (2005b). "From Cardinal to Ordinal Utility Theory: Darwin and Differential Capacity for Happiness." *American Journal of Economics and Sociology* 64(3): 851–879.

Poliakov, Léon. (1971). *Le mythe aryen: Essai sur les sources du racisme et des nationalismes*. Paris: Calmann-Lévy.

Quinlan, Sean (1998). "The Racial Imagery of Degeneration and Depopulation: Georges Vacher de Lapouge and 'Anthroposociology' in Fin-de-Siècle France." *History of European Ideas* 24(6): 393–413.

Ripley, William Z. (1900). "Ethnic Theories and Movements of Population: A Rejoinder." *Quarterly Journal of Economics* 14(3): 426–429.

Rutherford, Malcolm. (2005). "Comments on Four Papers on Economics and Human Heterogeneity." *American Journal of Economics and Sociology* 64(3): 881–886.

Schneider, William H. (1990). *Quality and Quantity: The Quest for Biological Regeneration in Twentieth-Century France*. Cambridge: Cambridge University Press.

Smith, Mark M. (2005). "Finding Deficiency: On Eugenics, Economics, and Certainty." *American Journal of Economics and Sociology* 64(3): 887–900.

Sorokin, Pitirim A. (1927). *Social Mobility*. New York: Harper and Brothers.

Spengler, J. J. (1944). "Pareto on Population." *Quarterly Journal of Economics* 59(1): 107–133.

Spindler, Michael. (2002). *Veblen and Modern America: Revolutionary Iconoclast*. London: Pluto Press.

Staehle, Hans. (1943). "Ability, Wages, and Income." *Review of Economic Statistics* 25(1): 77–87.

Stark W. (1961). "Natural and Social Selection." In *Darwinism and the Study of Society: A Centenary Symposium*. Ed. M. Banton. London: Tavistock Publications.

Stein, George J. (1988). "Biological Science and the Roots of Nazism: The Promotion of Racist Doctrines in the Name of Science." *American Scientist* 76: 50–58.

Stocking, George W., Jr. (1968). *Race, Culture, and Evolution: Essays in the History of Anthropology.* Chicago: University of Chicago Press.

Tilman, Rick. (1996). *The Intellectual Legacy of Thorstein Veblen: Unresolved Issues.* Westport, CT: Greenwood Press.

Veblen, Thorstein. (1898). "Why Is Economics Not an Evolutionary Science?" *Quarterly Journal of Economics* 12(4): 373–397.

——. (1904). *The Theory of Business Enterprise.* New York: Charles Scribner's Sons.

——. (1913a). "The Mutation Theory and the Blond Race." *Journal of Race Development* 3: 491–507. Rpt. in Veblen 1919.

——. (1913b). "The Blond Race and the Aryan Culture." *University of Missouri Bulletin,* Science Series 2: 39–57. Rpt. in Veblen 1919.

——. (1919). *The Place of Science in Modern Civilization and Other Essays.* New York: B.W.Huebsch.

——. ([1914] 1922). *The Instinct of Workmanship: And the State of the Industrial Arts.* New York: B.W. Huebsch.

Weintraub, E. Roy. (2003). "Prejudice and the History of Economics: A Minisymposium." *History of Political Economy* 35(4): 685–686.

Weiss, Sheila Faith. (1987). *Race Hygiene and National Efficiency: The Eugenics of Wilhelm Schallmayer.* Berkeley: University of California Press.

INDEX

for

SOCIAL INEQUALITY, ANALYTICAL EGALITARIANISM, AND THE MARCH TOWARD EUGENIC EXPLANATIONS IN THE SOCIAL SCIENCES